DISCRETE THOUGHTS

Essays on Mathematics, Science, and Philosophy

Mark Kac, Gian-Carlo Rota, and Jacob T. Schwartz

DISCRETE THOUGHTS

Essays on Mathematics, Science, and Philosophy

Birkhäuser

Boston · Basel · Stuttgart

Edited by Harry Newman
DISCRETE THOUGHTS is part of the series **Scientists of Our Time**
Series Editors: Gian-Carlo Rota and David Sharp

Library of Congress Cataloging in Publication Data

Kac, Mark.
Discrete thoughts.

Includes index.
1. Mathematics—Addresses, essays, lectures. 2. Science—Addresses, essays,
lectures. 3. Philosophy—Addresses, essays, lectures. I. Rota, Gian-Carlo,
1932-   . II. Schwartz, Jacob T. III. Newman, Harry. IV. Title. QA8.6.K33
1985 510 84-28212 ISBN 0-8176-3285-9

CIP-Kurztitelaufnahme der Deutschen Bibliothek

Discrete Thoughts : essays in mathematics,
science, and philosophy / Mark Kac, Gian-Carlo
Rota, and Jacob T. Schwartz. Ed. by Harry Newman.
– Boston ; Basel ; Stuttgart : Birkhäuser, 1985.

– ISBN 3-7643-3285-9

NE: Kac, Mark [Mitverf.]; Rota, Gian-Carlo [Mitverf.]; Schwartz, Jacob T.
[Mitverf.]; Newman, Harry [Hrsg.]

27;70

ISBN 0-8176-3285-9
ISBN 3-7643-3285-9

Printed in the United States of America

Dedicated to Nick Metropolis

TABLE OF CONTENTS

PREFACE

IN MATHEMATICS, as anywhere today, it is becoming more difficult to tell the truth. To be sure, our store of accurate facts is more plentiful now than it has ever been, and the minutest details of history are being thoroughly recorded. Scientists, PR-men and scholars vie with each other in publishing excruciatingly definitive accounts of all that happens on the natural, political and historical scenes.

Unfortunately, telling the truth is not quite the same thing as reciting a rosary of facts. José Ortega y Gasset, in an admirable lesson summarized by Antonio Machado's three-line poem, prophetically warned us that

> the reason people so often lie
> is that they lack imagination:
> they don't realize that the truth, too,
> is a matter of invention.

Sometime, in a future that is knocking at our door, we shall have to retrain ourselves or our children to properly tell the truth. The exercise will be particularly painful in mathematics. The enrapturing discoveries of our field systematically conceal, like footprints erased in the sand, the analogical train of thought that is the authentic life of mathematics. Shocking as it may be to a conservative logician, the day will come when currently vague concepts such as motivation and purpose will be made formal and accepted as constituents of a revamped logic, where they will at last be allotted the equal status they deserve, side-by-side with axioms and theorems. Until that day, however, the truths of mathematics will make only fleeting appearances, like shameful confessions whispered to a priest, to a psychiatrist, or to a wife.

In the nineteenth chapter of "The Betrothed," Manzoni describes as follows the one genuine moment in a conversation between astute Milanese diplomats: "It was as if, between acts in the performance of an opera, the curtain were to be raised too soon, and the spectators were given a glimpse of the half-dressed soprano screaming at the tenor."

Today, as in the last century, what rare glimpses of genuine expression we ever get will be found in offhand remarks hidden within an ephemeral essay we have allowed ourselves to write in a moment of weakness, or as a casual revelation inserted between the lines of some overdue book review. It takes chutzpah, and the nudging of an indulgent publisher, to bring anyone to the point of opening the drawer, grabbing the yellowed reprints of his occasional scribblings, and stringing them in the misleading linear order of chapters of what will optimistically be billed as a book. Yet, this is what three authors have jointly conspired to do here. They offer no excuses for their presumption, but they have not abandoned hope for the leniency that is granted to the accused after their fumbling attempts at telling the truth.

February 7, 1985
GIAN-CARLO ROTA

Acknowledgments

2   Reprinted by the permission of Boston University

3   Reprinted by the permission of the publisher, Stanford University Press, Stanford, California

4   Reprinted by the permission of the Applied Probability Turst, Sheffield, U.K.

6, 7, 9, 10   Reprinted by the permission of the MIT Press

8   Reprinted by the permission of the Oxford University Press

22   Reprinted by the permission of the Springer-Verlag, New York, New York

12   Reprinted by the permission of the American Mathematical Society

13   Reprinted by the permission of the Editor, Journal of the Franklin Institute

14   Reprinted by the permission of Martinus Nijhoff Publisher B.V., Netherlands

15   Reprinted by the permission of the World Research, Inc., San Diego, California

16   Reprinted by the permission of W.H. Freeman and Company, San Francisco

5, 18   Reprinted by the permission of the publisher, The Rockefeller University Press

20   Reprinted by the permission of the Editor, Philosophia Mathematica

1, 17,
21, 26   Reprinted by the permission of Academic Press, Orlando, Florida

4, 18   Reprinted by the permission of the Weizman Institute of Science, Israel

24   Reprinted from the Phenomenology Information Bulletin by the permission of the World Phenomenology Institute, Belmont, Mass.

25   Reprinted by the permission of the Association of American Colleges, Washington, D.C.

11   Reprinted by permission of Birkhäuser Boston, Inc., Cambridge, Mass.

CHAPTER ONE

Discrete Thoughts

OF SOME FIELDS it is difficult to tell whether they are sound or phony. Perhaps they are both. Perhaps the decision depends on the circumstances, and it changes with time. At any rate, it is not an objective fact like "the moon is made of green cheese." Some subjects start out with impeccable credentials, catastrophe theory for instance, and then turn out to resemble a three-dollar bill. Others, like dynamic programming, have to overcome a questionable background before they are reluctantly recognized to be substantial and useful. It is a tough world, even for the judgement-pronouncers.

It is much easier for a mathematician to read a physics book after the physics becomes obsolete, and that is in fact what usually happens. One wants to concentrate on the math, which is difficult to do if the physics is still under discussion. One then realizes how much good and new math physicists know without knowing it.

Gifted expositors of mathematics are rare, indeed rarer than successful researchers. It is unfortunate that they are not rewarded as they deserve, in our present idiotic pecking order.

On leafing through the collected papers of great mathematicians, one notices how few of their ideas have received adequate attention. It is like entering a hothouse and being struck by a species of flowers whose existence we did not even suspect.

The lack of real contact between mathematics and biology is either a tragedy, a scandal, or a challenge, it is hard to decide which.

Textbooks in algebraic geometry oscillate between Scylla and Charybdis: either they stick to the latest fad, thereby becoming accessible to the dozen-odd initiates, or they try to give a more balanced diet, thereby becoming estranged from the fold but available to a wide public.

Whatever one may think about engineers, one must admit they write clearly, to the point, and engagingly. They have realized long ago that if you wish the reader to follow rather then decipher, the linear deductive technique of exposition is the worst.

Lewis Carroll's "Symbolic Logic" is the *reductio ad absurdum* of Aristotelian logic. His syllogisms are not only good for a belly laugh; they actually display in the clarity of jest how abnormal and rare a form of reasoning a syllogism actually is. If anyone still thinks that Aristotelian logic has all but the most tenuous anchorage in reality, he will benefit from leafing through this amazing classic. After Lewis Carroll we can no longer teach the obsolete logic of yesteryear without a spellbreaking giggle.

First Johnny couldn't read, then Dick couldn't teach, now Harry doesn't know. It has been an acknowledged fact, since Poincaré pointed an accusing finger at the Twentieth Century, that much mathematics of our time has had negative overtones, and that a favored occupation of real-variable theorists and of set theoreticians, among others, has been the creation of those monsters that Francisco de Goya y Lucientes saw coming out of the dream of reason. But why harp on an accepted failing? The giants of Victorian certainty had their feet mired in hypocrisy and wishful thinking, and the hatchet job of negative mathematicians had a liberating influence that is just beginning to be felt. "Suppose a contradiction were to be found in the axioms of set theory. Do you seriously believe that that bridge would fall down?" asked Frank Ramsey of Ludwig Wittgenstein.

What if we were never able to certainly found set theory on a noncontradictory set of axioms? Maybe it is our Victorian ideas about the need for the certainty of axiomatization that are naive and unrealistic. Maybe the task ahead of us is to live rigorously with uncertainty, and bridges need not fall down in the attempt. *"Rien ne m'est sûr que la chose incertaine,"* was the motto of a Renaissance gentleman, reported by Montaigne. It may be the motto for our time too.

Mathematicians, like Proust and everyone else, are at their best when writing about their first love.

I

CHAPTER TWO

Mathematics: Tensions

FOR AS LONG AS WE CAN RELIABLY REACH into the past we find the development of mathematics intimately connected with the development of the whole of our civilization. For as long as we have a record of man's curiosity and quest for understanding we find mathematics cultivated and cherished, practiced and taught. Throughout the ages it has stood as an ultimate in rational thought and as a monument to man's desire to probe the workings of his own mind.

The strong urge to understand and to create mathematics has always been remarkable, considering that those who devoted their lives to the service of this aloof and elusive mistress could expect neither great material rewards nor widespread fame. Let me illustrate the kind of passion that mathematics could engender in the souls and minds of men: the Greeks left us two great areas of unsolved problems. The first was concerned with Euclid's Fifth Postulate and the other with whether certain geometric constructions (notably trisecting of an angle, doubling of a cube and squaring of a circle) could be performed by the sole use of a compass and an unmarked ruler.

The problem of the Fifth Postulate is purely logical. The Greeks felt uneasy about the necessity of assuming that through a point outside a given straight line one can pass one, and only one, parallel to that line. It was thought that perhaps this nearly self-evident statement could be logically derived from the remaining axioms and postulates which were considered to be more self-evident.

The problems of geometric constructions may appear even stranger when judged against the present-day, impatiently technological background. The Greeks as well as their successors were well aware that the constructions could be performed to any practical degree of accuracy with great ease. They were also aware that by allowing a wider range of instruments the constructions could be performed exactly. But they insisted on using the compass and the unmarked ruler as the sole instruments! It took nearly eighteen centuries of prodigiously futile and often anonymous efforts until the impossibility of the task was magnificently demonstrated during the past century.

It took almost equally long to settle the problem of the Fifth Postulate. And again pages of history are full of the struggle, Homeric in scope, against this seemingly invincible fortress. The magnificent obsession that pushed mathematics forward can be dramatized no better than by the life of the seventeenth-century monk, Saccheri, who in the austerity of his medieval cell labored so hard at proving the Fifth Postulate that he anticipated by more than a century some of the propositions of non-Euclidean geometry. And yet like many before him and many after him he died not knowing that his search was in vain, only because the laws of logic decreed that it had to be in vain. Perhaps if you think of Saccheri and others like him, you might forgive the pride with which the great nineteenth-century mathematician, Jacobi, declared that mathematics is practiced *"pour la gloire de l'esprit humain"* — for the glory of the human spirit. And indeed, is there a greater tribute to the human spirit than satisfaction of curiosity in its purest form?

The passionate pursuit of pure truth was not however the only force that lay behind the development and progress of mathematics. Mathematics, as we all know, helped man reach for the stars and explore the innermost secrets of the atom. Some of the greatest mathematical discoveries, calculus being the most notable among them, were brought about by man's quest for an understanding of the world about him. Even geometry, for ages the main source of intellectual purity in mathematics, originated, no doubt, in astronomy and geodesy. Because of this dichotomy of sources of inspiration mathematics has always played the roles of both the queen and the hand-maiden of science. This dual role proved to be a source of great strength. It permitted a healthy cross-fertilization of ideas, and it helped mathematics to steer a straight course between the dead ends of extreme pragmatism and empty and useless abstraction.

But the great advantages resulting from the duality of roles were often neither recognized nor acclaimed. Quite to the contrary. At various stages of our long and proud history the extreme "purists" in our midst have tried to free mathematics from its dual role. Let me quote from Plutarch, who in the *Life of Marcellus* speaks of the siege and defense of Syracuse:

> These machines he [Archimedes] had designed and contrived, not as matters of any importance, but as mere amusements in geometry; in compliance with King Hiero's desire and request, some little time before, that he should reduce to practice some part of his admirable speculation in science, and by accommodating the theoretic truth to sensation and ordinary use, bring it more within the appreciation of the people in general. Eudoxus and Archytas had been the first originators of this far-famed and highly-prized art of mechanics, which they employed as an elegant illustration of geometrical truths and as means of sustaining experimentally, to the satisfaction of the senses, conclusions too intricate for proof by words and diagrams. As, for example, to solve the problem, so often required in constructing geometrical figures, given the two extremes, to find the two mean lines of a proportion, both these mathematicians had recourse to the aid of instruments, adapting to their purpose certain curves and sections of lines. But what with Plato's indignation at it, and his invectives against it as mere corruption and annihilation of the one good in geometry, which was thus shamefully turning its back upon the unembodied objects of pure intelligence to recur to sensation, and to ask help (not to be obtained without base supervisions and depravation) from matter; so it was that mechanics came to be separated from geometry, and, repudiated and neglected by philosophers, took its place as a military art.

The followers of Plato are still with us, their indignation undiminished by passage of time, the futility of their efforts not fully recognized in spite of centuries of repudiation.

Our century, especially the last few decades, poured its own oil on troubled waters. With our predilection for labelling and classifying we have elevated the division of mathematics into "pure" and " applied" from the modest role of a semantic trick of dubious value to some kind of organizational reality. The college catalogues, the names of some learned societies and journals and a proliferation of committees, appointed to deal with a problem which should have never existed, all tend to add a kind of administrative sanction to a breach which may turn a turbulent but basically happy marriage into a hostile separation and divorce.

However, the internal conflict within mathematics is but a corollary of a wider and deeper conflict. Mathematics did not escape the centrifugal tendency which produced in more recent years what C. P. Snow so aptly calls the two cultures. The break in the unity of our culture was brought about, I think, by an unprecedented democratization of science. Spurred by complex sociological and political forces generated by hot and cold wars as well as the resulting technological expansion, science, almost overnight, changed from an avocation of few into a profession of many. As a result the lines of development became obscured and the delicate channels of communication overburdened to the point of breaking.

In a way, the problem of two cultures has always been with us. Unity was tenuously maintained through the past centuries because philosophy and mathematics somehow managed to bridge the gap. Rightly or wrongly, philosophy no longer fulfills this historic function. For one thing, it is almost universally distrusted, if not held in contempt, by a vast majority of modern practitioners of science. The position of mathematics has also been greatly weakened; first because it shared with the exact sciences a period of rapid democratization and second because this, in turn, widened the internal gulf within the discipline itself. In fact, unless something is done, "pure" mathematics will stake its fate on one culture, while "applied" mathematics will stake its fate on the other.

I have spoken of democratization of science and mathematics as the cause of a serious breach in our culture. I hasten to add that far from considering such a democratization evil or undesirable per se, I am happy to see the great flights of human imagination made available to a wide and ever-widening audience. The evil lies in our inability to deal with the process of democratization without sacrificing standards. On second thought, "standards" is not perhaps quite the right word. Integrity of motivation would come closer to what I am trying to convey. In an autobiographical sketch Einstein expresses his concern over this very problem in the following beautiful words:

> It is, in fact, nothing short of a miracle that the modern methods of instruction have not yet entirely strangled the holy curiosity of inquiry; for this delicate little plant, aside from stimulation, stands mainly in need of freedom; without this it goes to wreck and ruin without fail. It is a very grave mistake to think that the enjoyment of seeing and searching can be promoted by means of coercion and a sense of duty. To the contrary, I believe that it would be possible to rob even a healthy beast of prey of its voraciousness, if it were possible, with the aid of a whip, to force the beast to devour continuously, even when not hungry, especially if the food, handed out under such coercion, were to be selected accordingly.

The strong condemnation contained in this paragraph is unfortunately also applicable to mathematics.

By its nature and by its historical heritage mathematics lives on an interplay of ideas. The progress of mathematics and its vigor have always depended on the abstract helping the concrete and the concrete feeding the abstract. We cannot lose the awareness that mathematics is but a part of a great flow of ideas. To isolate mathematics and to divide it means, in the long run, to starve it and perhaps even to destroy it.

In recent years we have become much more preoccupied with streamlining and organizing our subject than with maintaining its overall vitality. If we are not careful, a great adventure of the mind will become yet another profession. Please, do not misunderstand me. The ancient professions of medicine, engineering or law are in no way inferior as disciplines to mathematics, physics or astronomy. But they are *different,* and the differences have been, since time immemorial, emphasized by vastly different educational approaches. The main purpose of professional education is development of *skills;* the main purpose of education in subjects like mathematics, physics or philosophy is development of *attitudes*. When I speak of professionalism in mathematics it is mainly this distinction that I have in mind. Our graduate schools are turning out ("producing" is the horrible word which keeps creeping up more and more) specialists in topology, algebraic geometry, differential equations, probability theory or what have you, and these in turn go on turning out more specialists. And there is something disconcerting and more than slightly depressing in the whole process. For if I may borrow an analogy, we no longer climb a mountain because it is there but because we were trained to climb mountains and because mountain climbing happens to be our profession.

It is the so-called "applied" mathematics that is usually blamed for professionalization of mathematics. The complex needs of our society have produced an unprecedented demand for "professional" mathematicians. Clearly, neither our industry nor our government looked for mathematicians of Saccheri's ilk. What was needed was competence, high-level competence to be sure, but not an obsession with a seemingly abstruse logical difficulty. In spite of this, it was realized that the whole of mathematics including its most "useless" branches ought to be supported. Support produced democratization and democratization in turn produced professionalism, ironically enough, even in the "purest" parts of mathematics.

Ever since Plato's indignation at Eudoxus and Archytas, the doctrine that applicability corrupts has been held by many leading mathematicians. In his charming book, *A Mathematician's Apology,* the late G. H. Hardy, one of the most eminent mathematicians of our era, took pride in the fact that nothing he had ever done could be conceivably applied to anything outside pure mathematics. Apart from the fact that this was not quite factually correct (I happen to know that a result of Hardy and Ramanujan in pure number theory was applied to a problem of distribution of energy levels in a nucleus), I find this attitude just as conducive to fostering professionalism as that of an incurably pragmatic "applied" mathematician who takes pride in not needing the "fancy abstractions" of his " pure" confrères. There is only one source of corruption in mathematics, and that is the motivation by anything except curiosity and desire for understanding and harmony. To attract disciples in terms of employment opportunities is already a sin. Not even the lofty terms of serving the needs of society and mankind are free from corrupting influences.

Mathematicians all over the world feel this very strongly but they often confuse purity of the subject with "purity" of subject matter. It is not corrupting either for the science or for the individual to be engaged in a problem which can be applied to physics, astronomy or even ballistics; what is corrupting is to become engaged in such a problem solely because it can be applied. But by the same token it is equally corrupting not to get interested in a problem because it may somehow prove "useful."

In summary, let me try to state my fears for the immediate future of mathematics. There is at present an enormous need for mathematical skills. The mathematical community has the responsibility to the society of which it is a part to help fill this need. It must however cope with this vastly difficult task without destroying the basic integrity of the subject. While serving reality we must not abandon the dream; while performing the task we must keep alive the passion. Can we do it?

The extreme purists in our midst urge division and separation. Like Plato, centuries before them, they see in applications of mathematics "mere corruption and annihilation of the one good in mathematics" and they want mathematics to return to "the unembodied objects of pure intelligence." In their horror of one kind of professionalism they are creating another — professional purism. Our extreme pragmatists preach division and separation on different grounds. I would view separation as a tragedy, and I consider its prevention to be a major challenge of the years to come. The two great streams of mathematical creativity are a tribute to the universality of the human genius. Each carries its own dreams and its own passions. Together they generate new dreams and new passions. Apart both may die — one in a kind of unembodied sterility of medieval scholasticism and the other as a part of military art.

The words above were written fifteen years ago.

These fifteen years have been a time of such strife and turbulence that it will take a long time before the memories dim and the scars heal. Universities suffered perhaps most, for their very existence as autonomous institutions dedicated to creating and propagating knowledge came into question. The effect however of this turmoil on mathematics (as well as on all scientific disciplines) was hardly detectable. Though many of its young practitioners manned the barricades, mathematics experienced a period of singular growth and achievement, and its trends and its tensions remained very much the same as before. Except that the tensions surfaced in the form of the controversy over "New Math."

I will say at the outset that I was and still am a strong opponent of "New Math." While I do not consider it to have been an unmitigated disaster (there are many who do), and while I think that some of its effects will in the long run prove to be beneficial, I still think that on the whole it was a mistake to force it upon elementary and secondary schools.

Capitalizing on the dismay over the Soviet sputnik, the proponents of educational reform attributed our loss of the space race to a faulty educational system and especially to a deficient and outmoded way in which mathematics was taught. The proposed solution was to teach it *the right way,* based on understanding the fundamental concepts and on minimizing the traditional role of drill and other dull routines. *The right way* rested on a much greater reliance on strict logic, rigor and axiomatics than had been customary in the past and it was much closer to the way those who chose mathematics as their life's work are taught the fundamentals of their subject. Let me illustrate with an example the kind of change "New Math" introduced:

For generations students taking elementary geometry were

subjected to a proof that an isosceles triangle is equiangular, i.e., if in a triangle ABC, AC = BC, then the angle A is equal (congruent) to the angle B. The usual proof consists in dropping the perpendicular from C to the base AB of the triangle, denoting the point at which the perpendicular intersects AB by D and noting that the right triangles ADC and BDC are congruent. The proof is as usual accompanied by a diagram, and naturally D is shown to lie *between* A and B, a fact which is essential in the proof.

Not until the second half of the nineteenth century has anyone questioned the validity of so obvious a proof. But the proof is incomplete because the usual axioms of Euclid do not deal with the concept of "betweenness," and a set of new axioms dealing with this concept (so-called axioms of order) had to be added to "legitimize" proofs like the one discussed above. If this strikes one as the height of pedantic hairsplitting let me hasten to say that it is not. To see that one is confronted with a real issue one must stop and ask oneself whether geometry can be taught to a computer. The answer to this question would have to be "no" if only Euclid's axioms were given to it, for whenever a need would arise to use the fact that a point lies between two other points, the computer would be helpless, not being able to "see" what is manifestly evident to the human eye.

A set of axioms so complete as to make even the computer happy was not proposed until 1895 (by David Hilbert) in a feat of deep and subtle thought which has had a profound influence on the subsequent development of mathematics. But this surely is not a sufficient reason for feeding it to the young. Well, the apostles of "New Math" decided that it was wrong to teach geometry without axioms of order and so into the textbooks they went.

Apart from its pedagogical unsoundness (my Cornell colleague, the late W. A. Hurwitz, used to say that in teaching on an elementary level one must tell the truth, nothing but the truth, but not the *whole* truth) the approach to the teaching of geometry (as well as of algebra and arithmetic) with such an emphasis on logical subtleties constituted a shockingly arbitrary decision to force upon the young a decidedly one-sided picture of mathematics. To me and to many of my colleagues mathematics is not just an austere, logical structure of forbidding purity, but also a vital, vibrant instrument for understanding the world, *including* the workings of our minds, and this aspect of mathematics was all but lost. Complete axiomatization, someone has rightly said, is an *obituary* of an idea, and Hilbert's great feat was, in a way only a magnificent necrology of geometry. Anyway, there are worse things than being wrong, and being dull and pedantic are surely among them.

The "New Math" controversy was nevertheless of great interest because it brought into the open the deep division within the mathematical community as to what mathematics is and how it should be taught. Too bad that elementary and high school teachers with only a limited understanding of the centuries-long tensions within mathematics had to become involved in the battle. Pity also the untold numbers of innocent children who never learned how to add or multiply fractions, while being subjected to elements of set theory (not very well, most of the time!) and frightened by the spectacle of errors which might result from confusing numbers with numerals. ("Never teach how not to commit errors which are not likely to be committed" is another sadly forgotten principle of sound pedagogy.)

Today passions generated by "New Math" have largely subsided, but the tensions within mathematics, though no longer on public display, are still with us. More so, in fact, than fifteen years ago, because there are so many more practitioners of this ancient art and, more importantly, because new applications of mathematics, not always sound, and frequently not wholly honest, keep multiplying at an unbelievable rate. The 1968 report of the Committee on Support of Research in Mathematical Sciences (COSRIMS) speaks of the unprecedented mathematization of our culture, and indeed there is hardly a corner of intellectual activity that has not, in one way or another, been touched by mathematics. Not, to repeat, always soundly, and frequently not with complete honesty, but touched nevertheless.

Of course, the computer has had a great deal to do with that, and if one thinks of Plato's indignation at Eudoxus' and Archytas' innocent instruments merely because they lacked the "purity" of the ruler and compass, his reaction to a computer would make the anger of Achilles appear barely detectable by comparison. And yet the computer has achieved a place of importance in mathematics. Its very existence has given rise to a whole new area of challenging problems, and the close connections between mathematical logic and the theory of computability have grown deeper and more intimate. It has also contributed its share to the tensions by sprouting a discipline of Computer Science which, having failed to find a congenial domicile within the traditional citadels of mathematics and, in spite of being to some extent "despised and neglected" (but not by philosophers!), refused to become merely a "part of military art," and is flourishing on its own in specially created academic departments.

Also, much of what goes under the name of Applied Mathematics has in a number of institutions become administratively separated from "Core Mathematics," a convenient though somewhat misleading alias for "pure" mathematics. Given the realities of academic life in the United States, administrative separations tend to persist, and lead ultimately to intellectual polarizations.

Thus, what I feared fifteen years ago has to an extent come to pass, but fortunately the overall vitality of mathematics continues unimpaired. The purest "pure," at its best, is far from withering in a kind of "unembodied sterility of medieval scholasticism," and its uncompromising pragmatic counterpart, again at its best, shows no sign of oblivion as "a part of military art." One begins even to see here and there signs of rapprochement, which makes one believe that left to ourselves we can look forward to an exciting and productive, though not always peaceful, future.

The danger, and it is a real danger, is that severe economic pressures will upset the delicate and precarious balance. In hard times, the applied, the " relevant" and the concrete win over the "pure" and the abstract. And yet, suppressing the latter trends would just as surely destroy mathematics as isolating it from the excitement and stimulation offered by the external world would have, had it been allowed to continue unchecked. That some unwise leaders have in the past promoted such an isolation and that, emulating Plato, they have "despised and neglected" those disciplines which failed to deal only with "unembodied objects of pure intelligence" is no reason to repay them in kind now that the glove may be on the other hand.

However, serious educational reforms, especially on the undergraduate level, will have to be introduced to undo some of the ills resulting from the long adherence to the mainly purist views of the mathematical establishment. Here we may as well imitate our arch rival, the Soviet Union, where mathematical education is almost the same as it was in the Czarist time. They somehow managed to beat us with the sputnik without any New Math and they manage to maintain mathematics on the highest level (certainly comparable to ours) while teaching it with due respect to all of its aspects, as well as to its intimate ties with sister disciplines, notably physics. If in a society so rigidly controlled by an inhumane and pragmatic regime mathematics can flourish, seemingly, at least, free from its internal tensions, then how much better we should be able to do in a free society, provided only that we can find the will.

1976

CHAPTER THREE

The Pernicious Influence of Mathematics on Science

I WISH TO MAKE USE of the license permitted to a lay speaker, in two regards: in the first place, to confine myself to the negative aspects of this role, leaving it to others to dwell on the amazing triumphs of the mathematical method; in the second place, to comment not only on physical science but also on social science, in which the characteristic inadequacies which I wish to discuss are more readily apparent.

Computer programmers often make a certain remark about computing machines, which may perhaps be taken as a complaint: that computing machines, with a perfect lack of discrimination, will do any foolish thing they are told to do. The reason for this lies of course in the narrow fixation of the computing machine "intelligence" upon the basely typographical details of its own perceptions — its inability to be guided by any large context. In a psychological description of the computer intelligence, three related adjectives push themselves forward: single-mindedness, literal-mindedness, simple-mindedness. Recognizing this, we should at the same time recognize that this single-mindedness, literal-mindedness, simple-mindedness also characterizes theoretical mathematics, though to a lesser extent.

It is a continual result of the fact that science tries to deal with reality that even the most precise sciences normally work with more or less ill-understood approximations toward which the scientist must maintain an appropriate skepticism. Thus, for instance, it may come as a shock to the mathematician to learn that the Schrodinger equation for the hydrogen atom, which he is able to solve only after a considerable effort of functional analysis and special function theory, is not a literally correct description of this atom, but only an approximation to a somewhat more correct equation taking account of spin, magnetic dipole, and relativistic effects; that this corrected equation is itself only an ill-understood approximation to an infinite set of quantum field-theoretical equations; and finally that the quantum field theory, besides diverging, neglects a myriad of strange-particle interactions whose strength and form are largely unknown. The physicist, looking at the original Schrodinger equation, learns to sense in it the presence of many invisible terms, integral, integrodifferential, perhaps even more complicated types of operators, in addition to the differential terms visible, and this sense inspires an entirely appropriate disregard for the purely technical features of the equation which he sees. This very healthy self-skepticism is foreign to the mathematical approach.

Mathematics must deal with well-defined situations. Thus, in its relations with science mathematics depends on an intellectual effort outside of mathematics for the crucial specification of the approximation which mathematics is to take literally. Give a mathematician a situation which is the least bit ill-defined — he will first of all make it well defined. Perhaps appropriately, but perhaps also inappropriately. The hydrogen atom illustrates this process nicely. The physicist asks: "What are the eigenfunctions of such-and-such a differential operator?" The mathematician replies: "The question as put is not well defined. First you must specify the linear space in which you wish to operate, then the precise domain of the operator as a subspace. Carrying all this out in the simplest way, we find the following result..." Whereupon the physicist may answer, much to the mathematician's chagrin: "Incidentally, I am not so much interested in the operator you have just analyzed as in the following operator, which has four or five additional small terms — how different is the analysis of this modified problem?" In the case just cited, one may perhaps consider that nothing much is lost, nothing at any rate but the vigor and wide sweep of the physicist's less formal attack. But, in other cases, the mathematician's habit of making definite his literal-mindedness may have more unfortunate consequences. The mathematician turns the scientist's theoretical assumptions, i.e., convenient points of analytical emphasis, into axioms, and then takes these axioms literally. This brings with it the danger that he may also persuade the scientist to take these axioms literally. The question, central to the scientific investigation but intensely disturbing in the mathematical context — what happens to all this if the axioms are relaxed? — is thereby put into shadow.

In this way, mathematics has often succeeded in proving, for instance, that the fundamental objects of the scientist's calculations do not exist. The sorry history of the δ-function should teach us the pitfalls of rigor. Used repeatedly by Heaviside in the last century, used constantly and systematically by physicists since the 1920's, this function remained for mathematicians a monstrosity and an amusing example of the physicists' naiveté — until it was realized that the δ-function was not literally a function but a generalized function. It is not hard to surmise that this history will be repeated for many of the notions of mathematical physics which are currently regarded as mathematically questionable. The physicist rightly dreads precise argument, since an argument which is only convincing if precise loses all its force if the assumptions upon which it is based are slightly changed, while an argument which is convincing though imprecise may well be stable under small perturbations of its underlying axioms.

The literal-mindedness of mathematics thus makes it essential, if mathematics is to be appropriately used in science, that the assumptions upon which mathematics is to elaborate be correctly chosen from a larger point of view, invisible to mathematics itself. The single-mindedness of mathematics reinforces this conclusion. Mathematics is able to deal successfully only with the simplest of situations, more precisely, with a complex situation only to the extent that rare good fortune makes this complex situation hinge upon a few dominant simple factors. Beyond the well-traversed path, mathematics loses its bearings in a jungle of unnamed special functions and impenetrable combinatorial particularities. Thus, the mathematical technique can only reach far if it starts from a point close to the simple essentials of a problem which has simple essentials. That form of wisdom which is the opposite of single-mindedness, the ability to keep many threads in hand, to draw for an argument from many disparate sources, is quite foreign to mathematics. This inability accounts for much of the difficulty which mathematics experiences in attempting to penetrate the social sciences. We may perhaps attempt a mathematical economics — but how difficult would be a mathematical history! Mathematics adjusts only with reluctance to the external, and vitally necessary, approximating of the scientists, and shudders each time a batch of small terms is cavalierly erased. Only with difficulty does it find its way to the scientist's ready grasp of the relative importance of many factors. Quite typically, science leaps ahead and mathematics plods behind.

Related to this deficiency of mathematics, and perhaps more productive of rueful consequence, is the simple-mindedness of mathematics — its willingness, like that of a computing machine, to elaborate upon any idea, however absurd; to dress scientific brilliancies and scientific absurdities alike in the impressive uniform of formulae and theorems. Unfortunately however, an absurdity in uniform is far more persuasive than an absurdity unclad. The very fact that a theory appears in mathematical form, that, for instance, a theory has provided the occasion for the application of a fixed-point theorem, or of a result about difference equations, somehow makes us more ready to take it seriously. And the mathematical-intellectual effort of applying the theorem fixes in us the particular point of view of the theory with which we deal, making us blind to whatever appears neither as a dependent nor as an independent parameter in its mathematical formulation. The result, perhaps most common in the social sciences, is bad theory with a mathematical passport. The present point is best established by reference to a few horrible examples. In so large and public a gathering, however, prudence dictates the avoidance of any possible *faux pas*. I confine myself, therefore, to the citation of a delightful passage from Keynes' *General Theory,* in which the issues before us are discussed with a characteristic wisdom and wit:

"It is the great fault of symbolic pseudomathematical methods of formalizing a system of economic analysis...that they expressly assume strict independence between the factors involved and lose all their cogency and authority if this hypothesis is disallowed; whereas, in ordinary discourse, where we are not blindly manipulating but know all the time what we are doing and what the words mean, we can keep 'at the back of our heads' the necessary reserves and qualifications and adjustments which we shall have to make later on, in a way in which we cannot keep complicated partial differentials 'at the back' of several pages of algebra which assume they all vanish. Too large a proportion of recent 'mathematical' economics are mere concoctions, as imprecise as the initial assumptions they rest on, which allow the author to lose sight of the complexities and interdependencies of the real world in a maze of pretentious and unhelpful symbols."

The intellectual attractiveness of a mathematical argument, as well as the considerable mental labor involved in following it, makes mathematics a powerful tool of intellectual prestidigitation — a glittering deception in which some are entrapped, and some, alas, entrappers. Thus, for instance, the delicious ingenuity of the Birkhoff ergodic theorem has created the general impression that it must play a central role in the foundations of statistical mechanics.[1] Let us examine this case carefully, and see. Mechanics tells us that the configuration of an isolated system is specified by choice of a point p in its phase surface, and that after t seconds a system initially in the configuration represented by p moves into the configuration represented by $M_t p$. The Birkhoff theorem tells us that if f is any numerical

---

[1]This dictum is promulgated, with a characteristically straight face, in Dunford-Schwartz, *Linear Operators,* Vol. I, Chap. 7.

function of the configuration p (and *if* the mechanical system is metrically transitive), the time average

$$\frac{1}{T}\int_0^T f(M_t p)\,dt$$

tends (as $T \to \infty$), to a certain constant; at any rate for all initial configurations p not lying in a set e in the phase surface whose measure $\mu(e)$ is zero; μ here is the (natural) Lebesgue measure in the phase surface. Thus, the familiar argument continues, we should not expect to observe a configuration in which the long-time average of such a function f is not close to its equilibrium value. Here I may conveniently use a bit of mathematical prestidigitation of the very sort to which I object, thus paradoxically making an argument serve the purpose of its own denunciation. Let $\nu(e)$ denote the probability of observing a configuration in the set e; the application of the Birkhoff theorem just made is then justified only if $\mu(e) = 0$ implies that $\nu(e) = 0$. If this is the case, a known result of measure theory tells us that $\nu(e)$ is extremely small wherever $\mu(e)$ is extremely small. Now the functions f of principal interest in statistical mechanics are those which, like the local pressure and density of a gas, come into equilibrium, i.e., those functions for which $f(M_t p)$ is constant for long periods of time and for almost all initial configurations p. As is evident by direct computation in simple cases, and as the Birkhoff theorem itself tells us in these cases in which it is applicable, this means that $f(p)$ is close to its equilibrium value except for a set e of configurations of very small measure μ. Thus, not the Birkhoff theorem but the simple and generally unstated hypothesis "$\mu(e) = 0$ implies $\nu(e) = 0$" necessary to make the Birkhoff theorem relevant in any sense at all tells us why we are apt to find $f(p)$ having its equilibrium value. The Birkhoff theorem in fact does us the service of establishing its own inability to be more than a questionably relevant superstructure upon this hypothesis.

The phenomenon to be observed here is that of an involved mathematical argument hiding the fact that we understand only poorly what it is based on. This shows, in sophisticated form, the manner in which mathematics, concentrating our attention, makes us blind to its own omissions — what I have already called the single-mindedness of mathematics. Typically, mathematics knows better what to do than why to do it. Probability theory is a famous example. An example which is perhaps of far greater significance is the quantum theory. The mathematical structure of operators in Hilbert space and unitary transformations is clear enough, as are certain features of the interpretation of this mathematics to give physical assertions, particularly assertions about general scattering experiments. But the larger question here, a systematic elaboration of the world-picture which quantum theory provides, is still unanswered. Philosophical questions of the deepest significance may well be involved. Here also, the mathematical formalism may be hiding as much as it reveals.

1962

CHAPTER FOUR

Statistics

THERE IS A SAYING, attributed to Niels Bohr but apparently an old Danish proverb, that it is difficult to predict, especially the future. Therefore I will not engage in this extraordinarily difficult and involved task, but will restrict myself to illustrating some points which might be relevant.

First of all, it is easy to give numerous examples of research activity which started in one direction and happened to produce results in quite a different area. For example, when Sadi Carnot was working so hard on the steam engine, he was only interested in improving it. In actuality, he ended up discovering a fundamental law of nature — a discovery which has led to phenomenal consequences. At my own institution, which was then the Rockefeller Institute, O.T. Avery was working with his collaborators during the last world war on an extremely applied problem concerned with pneumonia. In the process of this investigation he discovered that the carriers of genetic information were not proteins but nucleic acids. As you all know, this discovery revolutionized the whole field of genetics, although the original problem which led to it was a specific applied medical problem. So, one never knows: it is not the problem, or the name attached to it that is pertinent. What matters is the special combination of the men, the problem, the environment — in fact, exactly those things which no one can possibly predict.

So now I would like to speak a little bit on what statistics, with or without the adjective "mathematical," has meant to me. Perhaps I am not quite the right person to speak about it, because my connections with statistics are somewhat tangential; nevertheless, input from the outside can sometimes be useful.

My first exposure to statistics, although I did not realize that it was statistics, took place when I was 14 or 15 years of age. My class had at that time an extraordinary teacher of biology who gave us an outline of Darwin's theory and, in particular, explained how one of the claims of that theory, namely that individual characteristics are inheritable, was demolished by an experiment and a little bit of thinking. This was done by W. Johannsen, who published his results in 1909. There is a particularly vivid description of it in a book by Professors Cavalli-Sforza and Bodmer (p. 523 of [1]). Johannsen took a large number of beans, weighed them and constructed a histogram; the smooth curve fitted to this histogram was what my teacher introduced to us as the Quetelet curve. That was my first encounter with the normal distribution and the name Quetelet.

At this time I would like to make a small digression which has its own point to make. Quetelet was an extraordinarily interesting man, who was a student of Laplace and was the first to introduce statistical methodology into social science (even though he was trained as an astronomer); he was also the author of some early books in the area *(Lettres sur la Théorie des Probabilités,* 1846; *Physique Sociale,* 1st ed. 1835, 2nd ed. 1869; *Anthropométrie,* 1870). It might interest some of you to know that Quetelet was private tutor to the two princes of Saxe-Coburg, one of whom, Prince Albert, later became Queen Victoria's Consort. He was the first major governmental figure to try to introduce some kind of rational thinking into the operations of the government, and thus may be considered as the forefather of Operations Research. Now, the point of this digression is that if you look back at this connection between Astronomy and Operations Research, via Laplace, Quetelet and Prince Albert, you will realize the significance of the Danish proverb I quoted at the beginning.

Anyway, coming back to Johannsen, he argued that if all individual characteristics are inheritable, then if we take the small beans and plant them, take the large ones and plant them, and plot separately the two histograms for the progeny of the small and the large beans, then we should again obtain Quetelet curves, one centered around the mean weight of the small beans used as progenitors and the other around that of the large ones. Now, he did carry out such an experiment and did draw those histograms, and discovered that the two curves were almost identical with the original one. Actually, there was a slight shift to the left for the small ones and to the right for the large ones, so that by repeated selection one could separate the two populations. Of course, we now know that it is possible to distinguish between the genetic and the environmental factors, because the mean is controlled by the genetic factor while the variance is controlled by the environmental.

I have mentioned the above example because it illustrates a kind of unity of scientific thought: here was a basic problem in biology which was solved by a rather simple idea, which on closer analysis turns out to have an underlying mathematical foundation; and when one thinks some more about it, one is able to put a great deal of quantitative flesh on this extraordinarily interesting and impressive qualitative skeleton.

Another example of the same kind in which there is nothing quantitative or mathematical to begin with is connected with the great James Clerk Maxwell. I am sure all of you know that he invented a demon which has been named after him; but it is relatively few people who know in what connection and for what purpose the demon was invented. You will find an excellent account of this in a magnificent article by Martin J. Klein, a distinguished physicist and a distinguished historian of physics (pp. 84-86 of [2]); but since this article might not be readily available to all of you, I take the liberty of repeating some of its contents.

As a matter of fact, Maxwell invented the demon in a reply to a letter from his friend Tait, who was writing a textbook on thermodynamics, and who always submitted to Maxwell for criticism whatever he wrote. Maxwell wrote to him: "Any contributions I could make to that study are in the way of altering the point of view here and there for clearness or variety, and picking holes here and there to ensure strength and stability." Maxwell then proceeded to pick the following hole (remember this was in 1867): He suggested a conceivable way in which "if two things are in contact, the hotter *could* take heat from the colder without external agency," which would absolutely contradict the orthodox statement of the Second Law of Thermodynamics.

To quote from [2], "Maxwell considered a gas in a vessel divided into two sections, *A* and *B*, by a fixed diaphragm. The gas in *A* was assumed to be hotter than the gas in *B*, and Maxwell looked at the implications of this assumption from the molecular point of view. A higher temperature meant a higher average value of the kinetic energy of the gas molecules in A compared to those in *B*, which is now well known to every student who has taken the elementary course in physics or physical chemistry. But as Maxwell had shown some years earlier, each sample of a gas would necessarily contain molecules having velocities of all possible magnitudes, distributed according to a law known afterwards as Maxwellian, which is, as a matter of fact, the same probability law as that described by a Quetelet curve. "Now," wrote Maxwell, "conceive of a finite being who knows the paths and velocities of all the molecules by simple inspection but who can do no work except open and close a hole in a diaphragm by means of a slide without mass." This being is to be assigned to open the hole for an approaching molecule in *A* only if the molecule has a velocity less than the root mean square velocity of the molecules in *B*; it is to allow a molecule in *B* to pass through the hole into *A* only if its velocity exceeds the root mean square velocity of molecules in *A*. These two procedures are to be carried out alternately, so that the numbers of the molecules in *A* and *B* do not change. As a result of this procedure, however, "the energy in *A* is increased and that in *B* diminished; that is, the hot system has got hotter and the cold colder, and yet no work has been done; only the intelligence of a very observant and neat-fingered being has been employed." If we could only deal with the molecules directly and individually in the manner of this supposed being, we could violate the Second Law. "Only we can't," added Maxwell, "not being clever enough."

Some years later, in a letter to John William Strutt, better known as Lord Raleigh, he came even closer to what turned out to be one of the most significant breakthroughs in scientific thinking. "For," he said, again referring to his demon, "if there is any truth in the dynamical theory of gases, the different molecules in a gas of uniform temperature are moving with different velocities." "That was the essential thing," to quote again [2], and the demon only served to make its implications transparently clear. Maxwell even drew an explicit "moral" from his discussion: "The second law of thermodynamics has the same degree of truth as the statement that if you throw a tumblerful of water into the sea, you cannot get the same tumblerful of water out again," and you see here the birth, the real birth — before Boltzmann and certainly before Gibbs — of the statistical approach to problems in physics, which proved to be of such enormous impact and usefulness.

Notice there was not a single formula in all these letters; there was only the drawing of conclusions from the variability of the velocities, which again can be attributed to the molecular structure of matter, and it immediately made possible further progress by again putting quantitative meat on the qualitative bones.

The two examples which I have given above were selected with a view to illustrating the basic strength of statistics: in both cases the analysis was based on the variability or random fluctuations that were present in each. Now random fluctuations exist everywhere and are often treated as a nuisance — "the error term." It is statistical methodology that it able to extract such information as is contained in this variability; and as such statistics is not a branch of this-or-that well-established science, but a discipline in its own right, — an important part of scientific methodology. If you look at if from this point of view, then what is important is not whether it is "mathematical" statistics or "demographic" statistics or "applied" statistics, but how good it is as statistics.

Given that the subject of statistics cuts across inter-disciplinary boundaries, it is inevitable that it deals with problems which different research workers encounter in widely different areas. I want to give an example of this from my own experience. It goes back to the days of the last world war when one of the tasks which the Radiation Laboratory was charged with was the improvement of radar. Although most of the work in this connection was highly applied, there was also a theoretical group with which I was associated as a consultant. One problem that this group was interested in was to find out how the observer who is watching the radar scope reacts to what he sees on the scope. To simplify the problem, suppose that the observer on duty has been instructed to watch the scope diligently and whenever he sees a blip to ring a warning bell, because there might be an enemy airplane. Now the problem is: When is a blip a blip? That is to say, when is a blip due to a signal and when is it due to (random!) noise?

The following experiment was performed: An observer was placed in front of the scope and told to watch a certain spot; he was told that with a 50:50 chance a signal would be put on or not put on, and he was to say "yes" or "no," according as he saw something or did not see anything. Now, of course, when the signal was strong, there was no question and the observer was right one hundred per cent of the time; but then the strength of the signal was slowly decreased until the signal-to-noise ratio became one or a little less, and the observer's probability of error underwent a corresponding increase.

Although the observer did not know in which trials there was a signal and in which there was not, the people in charge knew and were able to estimate from the records the observer's probability of a correct answer. It was then possible to compare this actual performance with a theoretical ideal. The theory of the ideal observer, which was developed in 1944 by a group of workers including Arnold J. F. Siegert, used a line of reasoning which has been familiar to statisticians for a long time, but was new to us. (For a detailed description see the book by Lawson and Uhlenbeck [3].)

The reasoning proceeds as follows: If there is no signal, the displacement or deflection on the scope is a random variable with density f_0; if there is a signal, the density is f_1. In actual practice, they can be taken to be two normal densities with the same variance and means proportional to the strength of the signal. Now, "the ideal observer" knows f_0 and f_1 and is extraordinarily clever; and he sets out to construct a rule for when to say "yes" and when to say "no," the rule to be such as to minimize the probability of error. Now, we are on familiar ground, and the solution is obviously that given by a variant of the Neyman-Pearson theory, although the workers on the project did not know it as such.

Looking back at this experience, I wonder how many research workers there are in various fields all over the world, who are at this time struggling with problems whose solutions already exist in the statistical literature.

Anyway, there is a sequel to the above story. Many years later, I was consulting with an outfit involved with the space program, in particular with automatization of signal detection. The problem was to pre-teach an automaton so that it can learn to detect signals which it receives when floating around in outer space. In this connection it occurred to me to try to see whether an automaton exposed to the problem of signal detection could discover for itself the Neyman-Pearson theory empirically. To make a long story short, it turns out that in the very special case in which f_0 and f_1 are Gaussian, with the same variance it is possible to devise an automaton and a learning program which turns the automaton into an ideal observer. Briefly, the learning program is as follows: The automaton sets itself an arbitrary threshold, and it is instructed that whenever the signal received exceeds the threshold, it must say "yes," otherwise "no"; if it answers correctly, the threshold is maintained at the same level, otherwise it is moved right or left depending on the kind of error; and so the process continues. It turns out that what we have here is a random walk with an attraction toward the true threshold; and then one can show that in a certain sense there is convergence in probability to the true threshold. (For a more detailed discussion see [4], [5] and especially the excellent book [6]. In general, one obtains a threshold criterion which is not necessarily that of the ideal observer, contrary to the erroneous claim made in [5].)

When this result was published, it came to the attention of psychologists interested in learning theory (notably Dr. D. Dorfman and his collaborators), who proceeded to make various experiments and modifications, and to publish papers in which I was referred to — with the result that I acquired an undeserved reputation among learning theory psychologists. (For a brief discussion of the psychological background see [6], pp. 25-26, where other references may be found.) More seriously, the interesting aspect is that a train of thought, which had started many years earlier in an applied problem in one area, ended up later in theoretical investigations in a different direction because of the underlying statistical current. The other interesting aspect is the problem of constructing an automaton that performs as well as an ideal observer in more general situations. It is not an easy problem, and it should lead to interesting mathematics.

Finally, my last example is one in the opposite direction: a pure mathematical context in which statistical thought makes it possible to state new kinds of mathematical problems. Although I like the example immensely, I must restrict myself to a brief reference to it, since I have repeated it many times in print and otherwise. Now many of you know what is known as Descartes' rule of signs for estimating the number of positive real roots of a polynomial with real coefficients; namely, the number of positive real roots is never larger than the number of changes of sign in the sequence of coefficients. Now, we can ask a question: "How good is Descartes' rule?"

This, of course, is a vague question, since there is no definition of what is meant by "good" or "not good" in this case. Now, there is nothing wrong with vague questions; it is the combination of vague questions and vague answers that is bad. Many imprecisely stated questions have a tremendous amount of good science in them. In our case, in order to measure how good Descartes' rule of signs is, we necessarily must consider an ensemble of polynomials; and once you have an ensemble, then necessarily you have to deal with it statistically. Now, every polynomial is identified uniquely by the vector of its coefficients; however, since the roots are unaffected if all the coefficients are multiplied by the same non-zero number, therefore for our purpose, all polynomials are identified by points on the surface of the unit sphere. Assuming these to be uniformly distributed, we can then estimate the average value of the number of real roots for polynomials of a high degree. As I have stated earlier, I do not wish to repeat the details here, but merely draw attention to how, starting with a simple vague question about a pure mathematical problem, we have generated more mathematics by taking a statistical approach. (For some details see [7].)

In conclusion I would like to point out that these examples indicate the interplay between ideas from different domains that goes on all the time, and is in fact extremely valuable for the development of human knowledge; they illustrate the fact that everything is connected to everything else, and it is impossible to separate completely any intellectual endeavor from any other. In particular, as regards mathematics, you cannot separate it from its applications to the external world, and you cannot separate statistics from mathematics, or mathematical statistics from applied statistics. Thus, in so far as I am at all able to look into the future and identify desirable directions for us to take, they point towards (a) unification, and (b) communication (amongst ourselves and with the outside world).

Let us be connected with as many reservoirs of inspiration and understanding as possible; let us ask questions, even vague questions, and try to answer them precisely; let us not worry about "pure," about "applied," about "useful," about "relevant," and so on. Let us, in brief, try to do our best, and what survives will be determined by Nature's Law of Survival of the Fittest; and what is fittest will be determined by the next generation in the light of what has survived. So be it!

References

1. CAVALLI-SFORZA, L. L. AND BODMER, W. F. *The Genetics of Human Populations,* W. H. Freeman & Co., San Francisco, 1971.
2. KLEIN, M. J. "Maxwell, his demon and the second law of thermodynamics," *Amer. Scientist* 58, 1970, 84-97.
3. LAWSON, J. L. AND UHLENBECK, G. E. *Threshold Signals,* Vol. 24, Radiation Laboratory
   Series, McGraw-Hill, New York, 1950.
4. KAC, M. "A note on learning signal detection," *IRE Trans. on Information Theory*
   8, 1962, 126-128.
5. KAC, M. "Some mathematical models in science," *Science* 166, 1969, 695-699.
6. NORMAN, M. F. *Markov Processes and Learning Models,* Academic Press, New York, 1972
7. KAC, M. "Signal and noise problems," *Amer. Math. Monthly* 61, 1954, 23-26.

1975

Chapter Five

Statistics and Its History

Introduction

ONE DOES NOT THINK OF THE NINETEENTH CENTURY as a century of scientific revolutions but as a period of growth and consolidation, and in the main this view is right.

It is during the nineteenth century that the mechanics of Galileo and Newton achieves ultimate perfection in the work of Hamilton and Jacobi. Astronomy, liberated by the Copernican revolution a little more than three centuries back, goes from triumph to triumph, culminating in the feat of predicting by calculation the position of a hitherto unobserved eighth planet of our system. The theory of heat, still primitive in the early days of the century, flowers and comes to maturity through an inspired succession of remarkable ideas of Carnot, Clapeyron, Robert Mayer, Joule, Clausius, Kelvin, and our own Josiah Willard Gibbs. Last but not least, Maxwell creates his magnificent theory of electromagnetic phenomena, which brings new light to light itself.

But with the possible exception of Maxwell's theory, which in some ways was revolutionary and which carried in it seeds of yet greater revolutionary changes, there were no clearly visible breaks in continuity of progress and no radical assaults on tradition and orthodoxy.

And yet there was a revolution in the making — a revolution that ultimately changed science in a way as profound as any except perhaps the Copernican one.

Mechanics Dominates the Nineteenth Century

The nineteenth century is dominated by mechanics. Even the electromagnetic theory of Maxwell is couched in mechanistic terms. The downfall of the caloric theory of heat is brought about by proofs that mechanical energy and heat are equivalent, thus hinting that perhaps thermal phenomena too can be brought within the reach of mechanics.

Mechanics achieves almost the stature of geometry — a model of rigor, precision, and intellectual purity.

Mechanics is also complete; if we only knew the positions and velocities of all bodies today, we could, by solving the equations of motion, predict the future course of the world.

What a magnificent design for the universe!

True, there are little difficulties here and there. For example, to know the positions and velocities of bodies, one must measure them, and measurements are subject to error. But this is set aside as only a technical matter. We must merely keep perfecting the instruments, something that in principle should be entirely feasible. And so at the end of the century Albert A. Michelson sees the future of physics as being devoted to making more and more refined measurements and not much else.

Enter Thermal Science

By the late sixties, thermal science (or thermodynamics as it is now called) is in essence also complete. It rests safely and firmly on two laws, one embodying the aforementioned equivalence of heat and mechanical work and the other though of earlier origin becomes known as the second law.

It is a much subtler principle, and as formulated by Clausius, it reads: "Heat can never pass from a colder to a warmer body without some other change, connected therewith, occurring at the same time."

What is meant is that passage of heat from a colder to a

warmer body cannot happen spontaneously but requires external work to be performed on the system. The law as stated allows no exceptions. The "never" means absolutely never!

The two laws yield a harvest of results and consequences so rich and varied as to be almost beyond belief. Hardly a corner of science (and engineering!) remains untouched or uninfluenced, and when some years later physics begins its struggle with quanta, the second law is the one beacon of light on the dark and uncharted seas.

Only the fundamental problem remains, and that is to fit the second law into the mechanistic frame. It is in facing up to this problem that one runs into difficulties whose resolution precipitates a crisis.

Are There Atoms?

The laws of thermodynamics describe thermal properties of bodies without regard to the *structure* of matter.[1] To strict thermodynamicists like Ostwald and Mach, thermodynamics was a closed subject albeit incomplete to the extent that, for example, equations of state, which describe how pressures and densities of substances are related at constant temperatures, had to be taken from outside the theory.

Thus the familiar Boyle-Charles law

$$pV = RT,$$

relating the pressure p and volume V of an ideal gas at constant temperature T is an *additional* empirical law as far as thermodynamics is concerned.[2]

It seems clear that to *derive* equations of state one must have

---

[1] It is this seeming limitation that proves to be a source of strength in the early days of the quantum revolution. For while quantum theory radically changed our views on the structure of matter, the changes could never go far enough to violate either the first or the second law.

[2] On the other hand, it *follows* from the laws of thermodynamics that the internal energy of a gas obeying the Boyle-Charles law depends only on temperature (that is, is independent of the volume), an important fact found experimentally by Joule.

some idea of how matter is put together, and it is here that the atomistic view enters the picture.

The idea that matter is composed of tiny invisible atoms has its origins in antiquity, and it surfaces throughout history to be repeatedly discarded and nearly forgotten. It is a strange hypothesis with no evidence to support it and nothing but philosophical fancies speaking in its favor.

But with John Dalton (1766-1844) there comes a change. For Dalton makes a fundamental discovery that chemicals combine to make other chemicals only in precisely defined ratios. Thus 16 grams of oxygen will combine with 2 grams of hydrogen to form water, but in *any other ratio* either some oxygen or some hydrogen will be left over.

It is a difficult law to reconcile with the picture of matter being continuous, and Dalton proposes a simple and compelling explanation on the basis of an atomistic view.

It suffices only to assume that oxygen is composed of little units whose weight is sixteen times as great as the weight of similar little units which make up hydrogen, and that two units of hydrogen combine with one oxygen unit in order to "explain" completely the strictness of the 16:2 ratio.

The revival of the atomistic hypothesis by Dalton marks the beginning of modern chemistry, and from now on the atomistic view is never without adherents and partisans. There are, of course, detractors and opponents, and they are, in fact, in the majority during the nineteenth century, but the atomists are not easily silenced even if the opposition does include figures of the magnitude of Wilhelm Ostwald and Ernst Mach.

There is, by the way, an analogy in the ways in which the atomistic and the Copernican hypotheses entered science. The arguments for both were based on the simplicity of description they provided. Neither could at first be decided upon on the basis of irrefutable experimental or observational evidence. Both contradicted the "evidence" of the senses. Both were strongly, almost violently, rejected by the "establishment." Both when finally accepted opened great new vistas for science.

Can One Derive Laws of Thermodynamics from Those of Mechanics?

If one accepts the view that matter is composed of particles (atoms or molecules), then its behavior and properties should follow from the laws of mechanics. In particular, the second law should be derivable from these laws, provided one knows (or assumes) something about the forces between the constituent particles.

To derive the second law from the laws of mechanics is an ambitious and difficult task that Ludwig Boltzmann set himself as a goal a little over a hundred years ago.

In 1872 Boltzmann published a paper in which among others he used an assembly of hard spheres (small billiard balls) colliding elastically in a vessel as a model of a monoatomic gas and showed that a state of equilibrium will be reached in a way entirely consistent with the second law. Since he seemingly used only the laws of mechanics, he could claim that at last he had derived the second law from mechanics.

Boltzmann's memoir, one of the landmarks in the history of exact sciences, precipitated a long, sometimes bitter, controversy. Only a few years after its appearance, Loschmidt pointed out that the derivation could not be entirely right, for the laws of mechanics are time reversible, while the approach to equilibrium is unidirectional and shows what is called the "arrow of time." In defense of his thesis Boltzmann was forced to appeal to a statistical interpretation of his results, but his explanations were not fully understood, and other difficulties of a logical nature were yet to come.

Of these, a difficulty that became known as the recurrence paradox seemed decisive. It was raised by Zermelo (a student of Planck's who inherited his suspicions of the validity of Boltzmann's approach from his teacher), who invoked a theorem of Poincaré which implied that a mechanical system of the kind Boltzmann considered will show a quasi-periodic behavior, returning repeatedly to states arbitrarily close to the initial one unless the initial state was of a most exceptional character.[3] This was in such an obvious contradiction with the approach to equilibrium that Zermelo was quite ready to dismiss the whole mechanistic approach to thermodynamics.

Paradoxes Resolved by Introducing the Statistical Point of View

Boltzmann again invoked his statistical interpretation but again failed to convince the opposition. And yet that a statistical treatment can resolve the difficulties and the paradoxes becomes clear by considering an idealized (and artificial) model of temperature equalization which was proposed in 1907 by Paul and Tatiana Ehrenfest.[4]

Here is a description of the model which is taken from an article that I wrote for the *Scientific American* in 1964:[5]

> Consider two containers, A and B, with a large number of numbered balls in A and none in B. From a container filled with numbered slips of paper pick a numeral at random (say 6) and then transfer the ball marked with that number from container A to container B. Put the slip of paper back and go on playing the game this way, each time drawing at random a number between 1 and N (the total number of balls originally in container A) and moving the ball of that number from the container where it happens to be to the other container.
>
> It is intuitively clear that as long as there are many more balls in A than there are in B the probability of drawing a number that corresponds to a ball in A will be considerably higher than vice versa. Thus the flow of balls at first will certainly be strongly from A to B. As the drawings continue, the probability of finding the drawn number in A will change in a way that depends on the past drawings. This form of dependence of probability on past events is called a Markov

---

[3]In technical terms, these exceptional states form a set of measure zero in the set of all allowable states, the measure in question being determined by the laws of mechanics.

[4]Paul and Tatiana Ehrenfest coauthored in 1911 a fundamental article for the *Encyclopedie der Mathematischen Wissenschaften* in which they explained with great clarity and precision Boltzmann's views. The article is available in English translation (by M. J. Moravcsik) published by the Cornell University Press in 1959.

[5]"Probability." *Scientific American*, Vol. 211, No. 3, September 1964, pp. 92-108 (in particular p. 106).

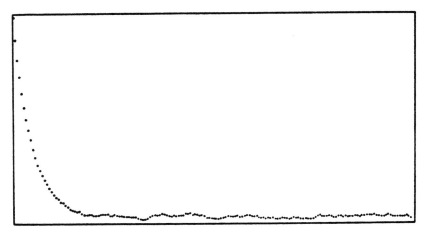

Figure 1. The Ehrenfest game played on a high-speed computer.

chain, and in the game we are considering, all pertinent facts can be explicitly and rigorously deduced. It turns out that, on an averaging basis, the number of balls in container A will indeed decrease at an exponential rate, as the thermodynamic theory predicts, until about half of the balls are in container B. But the calculation also shows that if the game is played long enough, then, with probability equal to 1, all the balls will eventually wind up back in container A, as Poincaré's theorem says!

How long, on the average, would it take to return to the initial state? The answer is 2^N drawings, which is a staggeringly large number even if the total number of balls (N) is as small as 100. This explains why behavior in nature, as we observe it, moves only in one direction instead of oscillating back and forth. The entire history of man is pitifully short compared with the time it would take for nature to reverse itself.

To test the theoretical calculations experimentally, the Ehrenfest game was played on a high-speed computer. It began with 16,384 "balls" in container A, and each run consisted of 200,000 drawings (which took less than two minutes on the computer). A curve was drawn showing the number of balls in A on the basis of the number recorded after every 1,000 drawings. As was to be expected, the curve of decline in the number of balls in A was almost perfectly exponential. After the number nearly reached the equilibrium level (that is, 8,192, or half the original number) the curve became wiggly, moving randomly up and down around that number. The wiggles were somewhat exaggerated by the vagaries of the machine itself, but they represented actual fluctuations that were bound to occur in the number of balls in A.

Behavior predicted by thermodynamics thus appears to be only the *average* behavior, but as long as the fluctuations are small and times of observation short compared with the "Poincaré cycles" (that is, times it takes for the system to come back near its initial state), it can be trusted with a high degree of confidence. Significant deviations from the average are so unlikely that they are hardly ever observed. And yet fluctuations are absolutely essential if the kinetic (molecular) and the thermodynamic views are to be reconciled.

What Price Atoms? Difficulties and Paradoxes

The price for the atomistic view was thus rejection of determinism and an acceptance of a statistical approach, and to the scientific establishment of the nineteenth century it was too high a price to pay. It was too high especially since no one has ever "seen" atoms or molecules, and there were no known violations of the second law. Why exchange classical simplicity and elegance of thermodynamics for a discipline full of difficulties, uncertainties, and paradoxes? ("Elegance," said Boltzmann bitterly, "should be left to shoemakers and tailors.")

But as the battle for the atomistic and therefore also nondeterministic view seemed on the verge of being lost, help came from a nearly forgotten discovery of an Irish botanist Robert Brown.

In 1827 Brown observed that small (but visible under a microscope) particles suspended in a liquid perform a peculiarly erratic motion. It was not however until 1905 when, independently of each other, A. Einstein[6] and M. Smoluchowski explained this phenomenon on the basis of kinetic theory. Brownian motion, they have shown, is a result of the particle being "kicked around" by the molecules of the surrounding liquid, and while the liquid appears to be in thermodynamic equilibrium on the macroscopic scale, it is on the microscopic scale in a state of disorganized thermal agitation. It is this state that contradicts strict thermodynamic behavior and calls for a probabilistic description.

The theory of Einstein and Smoluchowski not only explained Brownian motion but it made a number of predictions (for example, that the mean square displacement of a Brownian particle during a time t is proportional to t, the coefficient of proportionality known as the diffusion coefficient being simply related to the temperature and viscosity of the liquid)[7] which have been confirmed, within a few years, in a series of beautiful experiments by Jean Perrin.

Brownian motion has thus made thermal fluctuations visible and in this way the hitherto hypothetical atoms became real.

Fluctuations and Why the Sky Is Blue

Although the fluctuations are small, some of the effects they can produce are quite striking. Of these the most striking is the blueness of the sky.

On a sunny day when it is impossible to look directly at the sun, the color of the sky is determined by the *scattered light*. When a scatterer is small compared to the wavelength of incident, electromagnetic radiation is scattered with the intensity *inversely proportional to the fourth power of the wavelength* (the Rayleigh law), and hence in the case of visible light, it scatters much more light from the violet (short waves) end of the spectrum than from

---

[6]Einstein's first paper on the subject (published in *Annalen der Physik* in 1905) was entitled "On the Movement of Small Particles Suspended in a Stationary Liquid Demanded by the Molecular-Kinetic Theory of Heat" (I am quoting from an English translation of Einstein's papers on Brownian motion by A.D. Cowper which first appeared in 1926 and has been published by Dover Publications Inc. in 1956). When Einstein wrote this paper, he was unaware that the movement he so brilliantly analyzed had actually been observed.

His second paper "On the Theory of Brownian Movement," published in 1906, begins with the words:

"Soon after the appearance of my paper on the movements of particles suspended in liquids demanded by the molecular theory of heat, Siedentopf (of Jena) informed me that he and other physicists — in the first instance, Prof. Gouy (of Lyons) — had been convinced by direct observation that the so-called Brownian motion is caused by the irregular thermal movements of the molecules of the liquid."

[7]Even more importantly, the coefficient contains the so-called Boltzmann constant $k = R/N$, where R is the universal gas constant and N the "Avogadro number," that is, the universal number (6.02×10^{23}) of molecules on a mole of *any* substance. It was thus possible to determine N from Brownian motion experiments and compare the value thus obtained with determination based on different principles. The agreement was excellent.

the red end (long waves). But what are the small scatterers that do the job?

At first it was thought that tiny dust particles are the culprits, but this explanation had to be abandoned when it became clear that after a storm the air is much clearer of dust and yet the sky appears, if anything, even bluer.

The correct answer was finally provided by Smoluchowski and Einstein (about 1910), who pointed out that it was the individual atoms of gases of which air is composed that are the scatterers. But this is not enough for if the scatterers were arranged in a regular way (as in crystal), there would be hardly any scattering at all. It is only because the scatterers are distributed in a kind of random fashion that preferential scattering results in the blue color of the sky![8]

Another way of saying it is that density fluctuations provide the mechanism of light scattering.

If in a gas of average (number) density of ν molecules/cm^2 we consider a cube of side a cm, it will contain *on the average* νa^3 molecules. However, the actual number of molecules is not known precisely and is what one calls a *random variable*. In a nearly ideal gas the molecules are almost independent, and the deviation from the average can be shown to be of the order of $\sqrt{\nu a^3}$.[9]

The relative deviation is thus of the order

$$\frac{1}{\sqrt{\nu a^3}},$$

which for ν of the order of 2.5×10^{19} (corresponding to a gas at $0°$ under 1 atmosphere of pressure) and a of the order 5×10^{-5} cm (corresponding to the wavelength of visible light) is only a little less than one-tenth of 1 percent (0.1 percent).

---

[8]At sunset when we can look at the sun directly, the sky appeares reddish because now we do not see the scattered (blue) light and hence see the complementary color.

[9]This is a special case of what is technically known as the "weak law of large numbers." A more familiar consequence of this law is that in a series of n independent tosses of a coin the excess of heads over tails (or vice versa) is, for large n, of the order $\sqrt{n}$.

Visible light will therefore "see" air not as a homogeneous medium but as one having an irregularly granular structure. It is these deviations from regularity that are responsible for the scattering of light resulting in the blue color of the sky.

The End of the Revolution

In the preface to the second part of his monumental work *Lectures on Gas Theory*[10] Boltzmann wrote (in August 1898): "In my opinion it would be a great tragedy for science if the theory of gases were temporarily thrown into oblivion because of a momentary hostile attitude toward it…" And in the next paragraph he continued: "I am conscious of being only an individual struggling weakly against the stream of time." Only sixteen years later Marian Smoluchowski, the greatest Polish theoretician since Copernicus, delivers an invited address in Göttingen under the title "Gültigkeitsgrenzen des Zweiten Haupsatz der Wärmetheorie" (Limits of Validity of the Second Law of Thermodynamics) that only a few years earlier would have been considered subversive. In the address impressive experimental and theoretical evidence against the dogmatic view of the second law is carefully reviewed, and the *"never "* is replaced by "well, hardly ever."

The revolution had come to an end.

The status of the second law that for so long dominated the science of heat phenomena as it emerged from the struggle is best described in the words of Martin J. Klein.[11]

> Smoluchowski's brilliant analysis of the fluctuation phenomena showed that one could observe violations of almost all the usual statements of the second law by dealing with sufficiently small systems. The trend toward equilibrium, the increase of entropy, and so on, could not be taken as certainties. The one statement that could be upheld, even in the presence of fluctuations, was the impossibility of a perpetual motion of the second kind. No device could ever be made that would

---

[10]I am quoting from the English translation by Stephen G. Brush, University of California Press, 1964.

[11]Martin J. Klein, "Maxwell, His Demon, and the Second Law of Thermodynamics." *American Scientist,* Vol. 58, No. 1, 1970, pp. 84–97

use the existing fluctuations to convert heat completely into work on a macroscopic scale. For any such device would have to be constituted of molecules and would therefore itself be subject to the same chance fluctuations.

Or to use an analogy with games of chance which Smoluchowski himself liked to use: one can win occasionally, one can even amass a fortune, but one cannot design a *system* that would guarantee it.

1974

CHAPTER SIX

Combinatorics

COMBINATORIAL ANALYSIS — or, as it is coming to be called, combinatorial theory — is both the oldest and one of the least developed branches of mathematics. The reason for this apparent paradox will become clear toward the end of the present account.

The vast and ill-defined field of applied mathematics is rapidly coming to be divided into two clear-cut branches with little overlap. The first covers the varied offspring of what in the past century was called "analytical mechanics" or "rational mechanics," and includes such time-honored and distinguished endeavors as the mechanics of continua, the theory of elasticity, and geometric optics, as well as some modern offshoots such as plasmas, supersonic flow, and so on. This field is rapidly being transformed by the use of high-speed computers.

The second branch centers on what may be called "discrete phenomena" in both natural science and mathematics. The word "combinatorial," first used by the German philosopher and scientist G. W. Leibniz in a classic treatise, has been in general use since the seventeenth century. Combinatorial problems are found nowadays in increasing numbers in every branch of science, even in those where mathematics is rarely used. It is now becoming clear that, once the life sciences develop to the stage at which a mathematical apparatus becomes indispensable, their main support will come from combinatorial theory. This is already apparent in those branches of biology where the wealth of experimental data is gradually allowing the construction of successful theories, such as molecular biology and genetics. Physics itself, which has been the source of so much mathematical research, is now faced, in statistical mechanics and such fields as elementary particles, with difficult problems that will not be surmounted until entirely new theories of a combinatorial nature are developed to understand the discontinuous structure of the molecular and subatomic worlds.

To these stimuli we must again add the impact of high-speed computing. Here combinatorial theories are needed as an essential guide to the actual practice of computing. Furthermore, much interest in combinatorial problems has been stimulated by the possibility of testing on computers heretofore inaccessible hypotheses.

These symptoms alone should be sufficient to forecast an intensification of work in combinatorial theory. Another indication, perhaps a more important one, is the impulse from within mathematics toward the investigation of things combinatorial.

The earliest glimmers of mathematical understanding in civilized man were combinatorial. The most backward civilization, whenever it let fantasy roam as far as the world of numbers and geometric figures, would promptly come up with binomial coefficients, magic squares, or some rudimentary classification of solid polyhedra. Why then, given such ancient history, is combinatorial theory just now beginning to stir itself into a self-sustaining science? The reasons lie, we believe, in two very unusual circumstances.

The first is that combinatorial theory has been the mother of several of the more active branches of today's mathematics, which have become independent sometimes at the cost of a drastic narrowing of the range of problems to which they can be applied. The typical — and perhaps the most successful — case of this is algebraic topology (formerly known as combinatorial topology), which, from a status of little more than recreational mathematics in the nineteenth century, was raised to an independent geometric discipline by the French mathematician

Henri Poincaré, who displayed the amazing possibilities of topological reasoning in a series of memoirs written in the latter part of his life. Poincaré's message was taken up by several mathematicians, among whom were outstanding Americans such as Alexander, Lefschetz, Veblen, and Whitney. Homotopy theory, the central part of contemporary topology, stands today, together with quantum mechanics and relativity theory, as one of the great achievements in pure thought in this century, and the first that bears a peculiarly American imprint. The combinatorial problems that topology originally set out to solve are still largely unsolved. Nevertheless, algebraic topology has been unexpectedly successful in solving an impressive array of long-standing problems ranging over all mathematics. And its applications to physics have great promise.

What we have written of topology could be repeated about a number of other areas in mathematics. This brings us to the second reason why combinatorial theory has been aloof from the rest of mathematics (and that sometimes has pushed it closer to physics or theoretical chemistry). This is the extraordinary wealth of unsolved combinatorial problems, often of the utmost importance in applied science, going hand-in-hand with the extreme difficulty found in creating standard methods or theories leading to their solution. Yet relatively few men chose to work in combinatorial mathematics compared with the numbers active in any of the other branches of mathematics that have held the stage in recent years. One is reminded of a penetrating remark by the Spanish philosopher José Ortega y Gasset, who, in commenting upon the extraordinary achievements of physics, added that the adoption of advanced and accomplished techniques made possible "the use of idiots" in doing successful research work. While many scientists of today would probably shy away from such an extreme statement, it is nevertheless undeniable that research in one of the better developed branches of mathematics was often easier, especially for the beginner, than original work in a field like combinatorial theory, where sheer courage and a strong dose of talent of a very special kind are indispensable.

Thus, combinatorial theory has been slowed in its theoretical development by the very success of the few men who have solved some of the outstanding combinatorial problems of their day, for, just as the man of action feels little need to philosophize, so the successful problem-solver in mathematics feels little need for designing theories that would unify, and thereby enable the less talented worker to solve, problems of comparable and similar difficulty. But the sheer number and the rapidly increasing complexity of combinatorial problems have made this situation no longer tolerable. It is doubtful that one man alone could solve any of the major combinatorial problems of our day.

Challenging Problems

Fortunately, most combinatorial problems can be stated in everyday language. To give an idea of the present state of the field, we have selected a few of the many problems that are now being actively worked upon. Each of the problems has applications to physics, to theoretical chemistry, or to some of the more "businesslike" branches of discrete applied mathematics such as programming, scheduling, network theory, or mathematical economics.

1. *The Ising Problem*

A rectangular $(m \times n)$-grid is made up of unit squares, each colored either red or blue. How many different color patterns are there if the number of boundary edges between the red squares and the blue squares is prescribed?

This frivolous-sounding question happens to be equivalent to one of the problems most often worked upon in the field of statistical mechanics. The issue at stake is big: It is the explanation of the macroscopic behavior of matter on the basis of known facts at the molecular or atomic levels. The Ising problem, of which the above statement is one of many equivalent versions, is the simplest model that exhibits the macroscopic behavior expected from certain natural assumptions at the microscopic level.

A complete and rigorous solution of the problem was not achieved until recently, although the main ideas were initiated many years before. The three-dimensional analog of the Ising problem remains unsolved in spite of many attacks.

2. *Percolation Theory*

Consider an orchard of regularly arranged fruit trees. An infection is introduced on a few trees and spreads from one tree to an adjacent one with probability p. How many trees will be infected? Will the infection assume epidemic proportions and run through the whole orchard, leaving only isolated pockets of healthy trees? How far apart should the trees be spaced to ensure that p is so small that any outbreak is confined locally?

Consider a crystalline alloy of magnetic and nonmagnetic ions in proportions p to q. Adjacent magnetic ions interact, and so clusters of different sizes have different magnetic susceptibilities. If the magnetic ions are sufficiently concentrated, infinite clusters can form, and at a low enough temperature long-range ferromagnetic order can spread through the whole crystal. Below a certain density of magnetic ions, no such ordering can take place. What alloys of the two ions can serve as permanent magnets?

It takes a while to see that these two problems are instances of one and the same problem, which was brilliantly solved by Michael Fisher, a British physicist now at Cornell University. Fisher translated the problem into the language of the theory of graphs and developed a beautiful theory at the borderline between combinatorial theory and probability. This theory has now found application to a host of other problems. One of the main results of percolation theory is the existence of a critical probability p_e in every infinite graph G (satisfying certain conditions which we omit) that governs the formation of infinite clusters G. If the probability p of spread of the "epidemic" from a vertex of G to one of its nearest neighbors is smaller than the critical probability p_e, no infinite clusters will form, whereas if $p > p_e$, infinite clusters will form. Rules for computing the critical probability p_e were developed by Fisher from ingenious combinatorial arguments.

3. *The Number of Necklaces, and Polya's Problem*

Necklaces of n beads are to be made out of an infinite supply of beads in k different colors. How many distinctly different necklaces can be made?

This problem was solved quite a while ago, so much so that the priority is in dispute. Letting the number of different necklaces be $c(n,k)$, the formula is

$$c(n,k) \;=\; \frac{1}{n} \sum_{d|n} \phi(d)k^{n/d}.$$

Here, ϕ is a numerical function used in number theory, first introduced by Euler. Again, the problem as stated sounds rather frivolous and seems to be far removed from application. And yet, this formula can be used to solve a difficult problem in the theory of Lie algebras, which in turn has a deep effect on contemporary physics.

The problem of counting necklaces displays the typical difficulty of enumeration problems, which include a sizable number of combinatorial problems. This difficulty can be described as follows. A finite or infinite set S of objects is given, and to each object an integer n is attached — in the case of necklaces, the number of beads — in such a way that there are at most a finite number a_n of elements of S attached to each n. Furthermore, an equivalence relation is given on the set S — in this case, two necklaces are to be considered equivalent, or "the same," if they differ only by a rotation around their centers. The problem is to determine the number of equivalence classes, knowing only the integers a_n and as few combinatorial data as possible about the set S.

This problem was solved by the Hungarian-born mathematician George Polya (now at Stanford) in a famous memoir published in 1936. Polya gave an explicit formula for the solution, which has since been applied to the most disparate problems of enumeration in mathematics, physics, and chemistry (where, for example, the formula gives the number of isomers of a given molecule).

Polya's formula went a long way toward solving a great many

problems of enumeration, and is being applied almost daily to count more and more complicated sets of objects. It is nevertheless easy to give examples of important enumeration problems that have defied all efforts to this day, for instance the one described in the next paragraph.

4. *Nonself-intersecting Random Walk*

The problem is to give some formula for the number R_n of random walks of n steps that never cross the same vertex twice. A random walk on a flat rectangular grid consists of a sequence of steps one unit in length, taken at random either in the x- or the y-direction, with equal probability in each of the four directions. Very little is known about this problem, although physicists have amassed a sizable amount of numerical data. It is likely that this problem will be at least partly solved in the next few years, if interest in it stays alive.

5. *The Traveling Salesman Problem*

Following R. Gomory, who has done some of the deepest work on the subject, the problem can be described as follows. "A traveling salesman is interested in only one thing, money. He sets out to pass through a number of points, usually called cities, and then returns to his starting point. When he goes from the ith city to the jth city, he incurs a cost c_{ij}. His problem is to find that tour of all the points (cities) that minimizes the total cost."

This problem clearly illustrates the influence of computing in combinatorial theory. It is obvious that a solution exists, because there is only a finite number of possibilities. What is interesting, however, is to determine the minimum number $S(n)$ of steps, depending on the number n of cities, required to find the solution. (A "step" is defined as the most elementary operation a computer can perform.) If the number $S(n)$ grows too fast (for example if $S(n) = n!$) as the integer n increases, the problem can be considered unsolvable since no computer will be able to handle the solution for any but small values of n. By extremely ingenious arguments, it has been shown that $S(n)$ $\leq cn^2 2^n$, where c is constant, but it has not yet been shown that this is the best one can do.

Attempts to solve the traveling salesman problem and related problems of discrete minimization have led to a revival and a great development of the theory of polyhedra in spaces of n dimensions, which lay practically untouched — except for isolated results — since Archimedes. Recent work has created a field of unsuspected beauty and power, which is far from being exhausted. Strangely, the combinatorial study of polyhedra turns out to have a close connection with topology, which is not yet understood. It is related also to the theory behind linear programming and similar methods widely used in business and economics.

The idea we have sketched, of considering a problem $S(n)$ depending on an integer n as unsolvable if $S(n)$ grows too fast, occurs in much the same way in an entirely different context, namely, number theory. Current work on Hilbert's tenth problem (solving Diophantine equations in integers) relies on the same principle and uses similar techniques.

6. *The Coloring Problem*

This is one of the oldest combinatorial problems and one of the most difficult. It is significant because of all the work that has been done to solve it and the unexpected applications that this work has often had to other problems. The statement of the problem is deceptively simple: Can every planar map (every region is bounded by a polygon with straight sides) be colored with at most four colors, so that no two adjacent regions are assigned the same color?

Most of the early attempts at solving this problem (up to around 1930) were based on direct attack, and they not only failed but did not even contribute any useful mathematics. Thanks to the initiative of H. Whitney of the Institute for Advanced Study and largely to the work of W. T. Tutte (English-Canadian) a new and highly indirect approach to the coloring problem is being developed, called "combinatorial geometry" (or sometimes the theory of matroids). This is the first theory of a general character that has been completely successful in understanding a variety of combinatorial problems. The theory is based on a generalization of Kirchhoff's laws of circuit theory in a completely unforeseen — and untopological — direction. The basic notion is a closure relation with the MacLane-Steinitz exchange property. The exchange property is a closure relation, $A \rightarrow \bar{A}$, defined on all subsets A of a set S such that, if x and y are elements of S and $x \in \overline{A \cup y}$ but $x \notin \bar{A}$, then $y \in \overline{A \cup x}$. In general, one does not have $\overline{A \cup B} = \bar{A} \cup \bar{B}$, so that the resulting structure, called a combinatorial geometry, is not a topological space. The theory bears curious analogies with both point-set topology and linear algebra and lies a little deeper than either of them.

The most striking advance in the coloring problem is a theorem due to Whitney. To state it, we require the notion of "planar graph," which is a collection of points in the plane, called vertices, and nonoverlapping straight-line segments, called edges, each of them joining a pair of vertices. Every planar graph is the boundary of a *map* dividing the plane into *regions*. Whitney makes the following assumptions about the planar graph and the associated map: (a) Exactly three boundary edges meet at each vertex; (b) no pair of regions, taken together with any boundary edges separating them, forms a multiply connected region; (c) no three regions, taken together with any boundary edges separating them, form a multiply connected region. Under these assumptions, Whitney concludes that it is possible to draw a closed curve that passes through each region of the map once and only once. Whitney's theorem has found many applications since it was discovered.

7. *The Pigeonhole Principle and Ramsey's Theorem*

We cannot conclude this brief list of combinatorial problems without giving a typical example of combinatorial argument. We have chosen a little-known theorem of great beauty, whose short proof we shall give in its entirety. The lay reader who can follow the proof on first reading will have good reason to consider himself combinatorially inclined.

THEOREM: Given a sequence of $(n^2 + 1)$ distinct integers, it is possible to find a sequence of $(n + 1)$ entries which is either increasing or decreasing.

Before embarking upon the proof, let us see some examples. For $n = 1$, we have $n^2 + 1 = 2$ and $n + 1 = 2$; the conclusion is trivial since a sequence of two integers is always either increasing or decreasing. Let $n = 2$, so that $n^2 + 1 = 5$ and $n + 1 = 3$, and say the integers are 1,2,3,4,5. The theorem states that no matter how these integers are arranged, it is possible to pick out a string of at least three (not necessarily consecutive) integers that are either increasing or decreasing, for example,

$$1\ 2\ 3\ 4\ 5.$$

The subsequence 1 2 3 will do (it is increasing). Acutally, in this case every subsequence of three elements is increasing. Another example is

$$3\ 5\ 4\ 2\ 1.$$

Here all increasing subsequences, such as 3 4 and 3 5, have at most two integers. There is, however, a wealth of decreasing subsequences of three (or more) integers such as 5 4 2, 5 2 1.

One last example is

$$5\ 1\ 3\ 4\ 2.$$

Here there is one increasing subsequence with three integers, namely 1 3 4, and there are two decreasing subsequences with three integers, namely 5 3 2 and 5 4 2; hence, the statement of the theorem is again confirmed.

Proceeding in this way, we could eventually verify the statement for all permutations of five integers. There are altogether $5! = 120$ possibilities. For $n = 3$, we have to take $n^2 + 1 = 10$ integers, and the amount of work to be done to verify the conjecture case by case is overwhelming since the possibilities total $10! = 3,628,800$. We begin to see that an argument of an altogether different kind is needed if we are to establish the conclusion for all positive integers n.

The proof goes as follows. Let the sequence of integers (in the given order) be

$$a_1, \ a_2, \ a_3 \ ,\ldots, \ a_{n^2+1} \tag{1}$$

We are to find a subsequence of Sequence 1, which we shall label

$$a_{i_1}, \ a_{i_2}, \ \ldots, \ a_{i_{n+1}},$$

where the entries are taken in the same order as Sequence 1 but with one of the following two properties: either

$$a_{i_1} \leqq a_{i_2} \leqq \cdots \leqq a_{i_{n+1}}, \tag{2}$$

or

$$a_{i_1} \geqq a_{i_2} \geqq \cdots \geqq a_{i_{n+1}}, \tag{3}$$

The argument is based on a *reductio ad absurdum*. Suppose that there is no subsequence of the type of Sequence 2, that is, no increasing subsequence of $(n+1)$ or more entries. Our argument will then lead to the conclusion that, under this assumption, there must be a sequence of the type of Sequence 3, that is, a decreasing sequence with $(n+1)$ entries.

Choose an arbitrary entry a_i of Sequence 1, and consider all increasing subsequences of Sequence 1 whose first element is a_i. Among these, there will be one with a maximum number of entries. Say this number is l ($=$ length). Under our additional hypothesis, the number l can be 1, 2, 3,..., or n, *but not $n+1$* or any larger integer.

We have, therefore, associated to each entry a_i of Sequence 4 an integer l between 1 and n; for example, $l = 1$ if all subsequences of two or more integers starting with a_i are decreasing. We come now to the crucial part of the argument. Let $F(l)$ be the number of entries of Sequence 1 with which we have associated the integer l, by the procedure just described. Then

$$F(1) + F(2) + F(3) + \cdots + F(n) = n^2 + 1. \tag{4}$$

Identity 4 is just another way of saying that with each one of the $(n^2 + 1)$ entries, a_i of Sequence 1 we have associated a number between 1 and l. We claim that *at least one* of the summands on the left-hand side of Identity 4 must be an integer greater than or equal to $n + 1$. For if this were not so, then we should have

$$F(1) \leqq n, \ F(2) \leqq n, ..., F(n) \leqq n.$$

Adding all these n inequalities, we should have

$$F(1) + F(2) + \cdots + F(n) \leqq \underbrace{n + n + \cdots + n}_{n \text{ times}} = n^2,$$

and this contradicts Identity 4, since $n^2 < n^2 + 1$. Therefore, one of the summands on the left-hand side of Identity 4 must be at least $n + 1$. Say this is the lth summand:

$$F(l) \geqq n + 1.$$

We now go back to Sequence 1 and see what this conclusion means. We have found $(n + 1)$ entries of Sequence 1, call them (in the given order)

$$a_{i_1}, a_{i_2}, ..., a_{i_{n+1}} \tag{5}$$

with the property that each one of these entries is the beginning entry of an increasing subsequence of l entries of Sequence 1 but is not the beginning entry of any longer subsequence of Sequence 1.

From this we can immediately conclude that Sequence 5 is a *decreasing* sequence. Let us prove, for example, that $a_{i_1} > a_{i_2}$. If this were not true, then we should have $a_{i_1} < a_{i_2}$. The entry a_{i_2} is the beginning entry of an increasing subsequence of Sequence 1 containing exactly l entries. It would follow that a_{i_1} would be the beginning entry of a sequence of $(l + 1)$ entries, namely a_{i_1} itself followed by the sequence of l entries starting with a_{i_2}. But this contradicts our choice of a_{i_1}. We conclude that $a_{i_1} > a_{i_2}$. In the same way, we can show that $a_{i_2} > a_{i_3}$, etc., and complete the proof that Sequence 5 is decreasing and, with it, the proof of the theorem.

Looking over the preceding proof, we see that the crucial step can be restated as follows: If a set of $(n^2 + 1)$ objects is partitioned into n or fewer blocks, at least one block shall contain $(n + 1)$ or more objects or, more generally, if a set of n objects is partitioned into k blocks and $n > k$, at least one block shall contain two or more objects. This statement, generally known as the "pigeonhole" principle, has rendered good service to mathematics. Although the statement of the pigeonhole principle is evident, nevertheless the applications of it are often startling. The reason for this is that the principle asserts that an object having a certain property exists, without giving us a means for finding such an object; however, the mere existence of such an object allows us to draw concrete conclusions, as in the theorem just proved.

Some time ago, the British mathematician and philosopher F. P. Ramsey obtained a deep generalization of the pigeonhole principle, which we shall now state in one of its forms. Let S be an infinite set, and let $P_l(S)$ be the family of all finite subsets of S containing l elements. Partition $P_l(S)$ into k blocks, say $B_1, B_2, \ldots, B_k$; in other words, every l-element subset of S is assigned to one and only one of the blocks B_i for $1 \le i \le k$. Then there exists an infinite subset $R \subset S$ with the property that $P_l(R)$ is contained in one block, say $P_l(R) \subset B_i$ for some i, where $1 \le i \le k$; in other words, there exists an infinite subset R of S with the property that all subsets of R containing l elements are contained in one and the same of the B_i.

The Coming Explosion

It now seems that both physics and mathematics, as well as those life sciences that aspire to becoming mathematical, are conspiring to make further work in combinatorial theory a necessary condition for progress. For this and other reasons, some of which we have stated, the next few years will probably witness an explosion of combinatorial activity, and the mathematics of the discrete will come to occupy a position at least equal to that of the applied mathematics of continua, in university curricula as well as in the importance of research. Already in the past years, the amount of research in combinatorial theory has grown to the point that several specialized journals are being published. In the last few years, several textbooks and mongraphs in the subject have been published, and several more are now in print.

Before concluding this brief survey, we shall list the main subjects in which current work in combinatorial theory is being done. They are the following:

1. *Enumerative Analysis,* concerned largely with problems of efficient counting of (in general, infinite) sets of objects like chemical compounds, subatomic structures, simplicial complexes subject to various restrictions, finite algebraic structures, various probabilistic structures such as runs, queues, permutations with restricted position, and so on.

2. *Finite Geometries and Block Designs.* The work centers on the construction of finite projective planes and closely related structures, such as Hadamard matrices. The techniques used at present are largely borrowed from number theory. Thanks to modern computers, which allowed the testing of reasonable hypotheses, this subject has made great strides in recent years. It has significant applications to statistics and to coding theory.

3. *Applications to Logic.* The development of decision theory has forced logicians to make wide use of combinatorial methods.

4. *Statistical Mechanics.* This is one of the oldest and most active sources of combinatorial work. Some of the best work in combinatorial theory in the last twenty years has been done by physicists or applied mathematicians working in this field, for example in the Ising problem. Close connections with number theory, through the common medium of combinatorial theory, have been recently noticed, and it is very likely that the interaction of the two fields will produce striking results in the near future.

In conclusion, we should like to caution the reader who might gather the idea that combinatorial theory is limited to the study of finite sets. An infinite class of finite sets is no longer a finite set, and infinity has a way of getting into the most finite of considerations. Nowhere more than in combinatorial theory do we see the fallacy of Kronecker's well-known saying that "God created the integers; everything else is man-made." A more accurate description might be: "God created infinity, and man, unable to understand infinity, had to invent finite sets." In the ever-present interaction of finite and infinite lies the fascination of all things combinatorial.

1969

CHAPTER SEVEN

Computer Science

COMPUTER SCIENCE, a new addition to the fraternity of sciences, confronts its older brothers, mathematics and engineering, with an adolescent brashness born of rapid, confident growth and perhaps also of enthusiastic inexperience. At this stage, computer science consists less of established principles than of nascent abilities. It will, therefore, be the aim of this essay to sketch the major new possibilities and goals implicit in the daily flux of technique.

The power of computers is growing impressively. A few years ago, they carried out thousands of instructions per second. Today's models are capable of tens and hundreds of millions of instructions per second. This rapid advance is expanding the unique ability of the computer: its power to deal successfully with the irreducibly complex — that is, situations and processes that can be specified adequately only by very large amounts of information.

Without the high-precision and high-speed tool that is the computer, a process that is irreducibly complex in the above sense is unusable (and hardly conceivable) for two fundamental reasons. First, the accumulation and verification of great volumes of highly accurate interdependent information is not practical. For instance, because of inevitable human error in their preparation, completely accurate mathematical tables did not exist before computers became available. Second, since complex processes demand long sequences of elementary steps, no really complex process can actually be carried through without computers.

These two obstacles present themselves at so fundamental a level in any intellectual investigation that until the development of computers the possibility of dealing successfully with the complex itself was never really envisaged. Perhaps the most successful substitute for such a possibility, as well as the nearest approach to it, came in mathematics. Mathematics provided simplification in a different way: by furnishing intellectual techniques of conceptualizing abstraction and developing symbolic and nonsymbolic languages to facilitate the manipulation and application of concepts. Utilizing these techniques, mathematics has developed an impressive ability, not really to deal directly with the complex, but rather to hunt out ways to reduce complex content to principles that are briefly statable and transparent. To find the simple in the complex, the finite in the infinite — that is not a bad description of the aim and essence of mathematics.

Since the linguistic-conceptual procedure of mathematics is the most powerful method for helping the unaided intellect deal with precisely defined situations, the concepts and language of mathematics have been profitably adopted by many sciences. Beyond the individual strong points occupied by mathematics and marked out by its famous theorems and definitions lies a jungle of combinatorial particularities into which mathematics has always been unable to advance. Computer science is beginning to blaze trails into this zone of complexity. To this end, however, methods are required which, while they include some of the classical devices of mathematics, go beyond these in certain characteristic respects.

In this quest for simplification, mathematics stands to computer science as diamond mining to coal mining. The former is a search for gems. Although it may involve the preliminary handling of masses of raw material, it culminates in an exquisite item free of dross. The latter is permanently involved with bulldozing large masses of ore — extremely useful bulk material.

It is necessarily a social rather than an individual effort. Mathematics can always fix its attention on succinct concepts and theorems. Computer science can expect, even after equally determined efforts toward simplification, only to build sprawling procedures, which require painstaking and extensive descriptive mapping if they are to be preserved from dusty chaos.

In its initial efforts, information science has of course followed the easiest path, emphasizing those processes leading to results that are perhaps too complex for direct analytic treatment but can still be attained by use of relatively simple algorithms iteratively applied. These processes are complex only in the limited sense that carrying them out requires long sequences of simple steps. Most of the calculations, data analyses, simulations, and so forth, which are the common coin of present-day computer work, are large-scale repetitions of this kind. Elaborate programs of the types of language translators, linguistic analyzers, and theorem provers lie more definitely within the borders of the newly opened region of complexity into which computer science will progress as its equipment and technique grow stronger.

Technological Past and Perspectives

The modern computer is, of course, a *stored program* computer. That is, it includes a mass of recorded information, or *memory,* stored temporarily on any medium the engineer finds attractive, for example, magnetic tape, wired arrays of small circular magnets, punched cards, punched paper tape, celluloid strips. It also has one or more *processing* units, whereby segments of stored information are selected and caused to interact. The new information produced by this interaction is then again stored in memory, a procedure that allows successive steps of the same form to be taken.

Three separate segments of information are generally involved in each elementary interaction. Two of these are usually treated as coded representations of numbers to be combined by arithmetic or logical operations. The third is treated as a coded indication of the memory location from which these operands are to be drawn and of the operation to be performed: addition, multiplication, comparison, and so on. Such an elementary combination, or *instruction execution,* produces not only a direct result but also an indication of the place in memory where the result is to be recorded and an indication of the place where the next instruction is to be found. The power of a computer is roughly proportional to the rate at which it executes these elementary combinations.

A number of major recent technological developments have combined to advance computer power rapidly in the last decade.

1. The development of commercially available ultra-high-speed electronic circuits has made it practical to execute the elementary instructions in tenths or even hundredths of millionths of a second. The high reliability of transistors has made it possible to keep enormous assemblies of parts functioning simultaneously for extended periods. The development of "integrated" microelectronics, built up by the diffusion of electric catalysts into silicon or germanium chips, has reduced the size of computers, permitting signals to pass between their extremities in short periods. Computers now in design will be able to coordinate and to carry out a multiplication, a pair of additions, and a logical operation in the time it takes light to travel 15 feet, sustaining sequential instruction rates of 200 million instructions per second.

2. The development of manufacturing technique has reduced the cost of storing information in rapidly accessible form. Increasing sophistication in the over-all configuration of computer systems has made possible the effective and economic combination of available information storage media. The large computer system of the near future will have a hierarchy of rapidly accessible memories. A few hundred digits will be stored directly in high-speed electronic circuits, from which they may be read in some hundredths of a millionth of a second. Half a million to a few million digits in bulk electronic storage will be readable in well under 1/10 of a millionth of a second. And as many as 10-200 million digits in lower-grade magnetic storage will be accessible in 1/4 of a millionth of a second or less. Such

systems will also incorporate a variety of newly perfected bulk information storage devices, ranging through juke-box-like magnetic disk cabinets with fixed or movable recording arms, storing some hundreds of millions of digits accessible in tenths or hundredths of a second, plastic-strip information files storing billions of digits, and elaborate photographic microdot cabinets, in which trillions of digits can be stored for access in a few seconds. Such systems will constitute very impressive information-processing factories, in which an information base comparable in size to the Library of Congress can be processed at the rate of hundreds of millions of instructions per second.

3.   Increased sophistication in computer design has yielded substantially increased performance from given quantities of electronic circuitry. By artfully overlapping the execution of as many instructions as possible, a large part of the circuitry comprising a computer may be kept steadily in action. Ingenious techniques have been devised for the dynamic movement of data between the various storage media available to a computer to permit more effective use of the memory. Special arrangements permitting the computer to hold many tasks in immediate readiness, so that calculation may proceed on one while data are being fetched for others, also contribute to the productivity of a given computer complex.

4.   The continued rapid improvement of manufacturing technique and electronic reliability should soon make possible the construction of considerably larger circuit assemblages. These larger computers could, by allowing more parts of a logical process to proceed in parallel, attain very greatly increased instruction rates — up to several thousand million instructions per second. Improved manufacturing techniques should also make possible considerable increases in the capacities of fast computer memories and should eventually make possible the construction of memories that are themselves capable of some degree of logical activity.

5.   The availability of faster computers supplied with considerably increased banks of memory aids significantly in simplifying programming. In programming an algorithm to run on

a small slow computer, considerable effort is needed to compress the code and increase its efficiency. "Hand coding," reflecting this necessity, is more complex than mechanically generated codes would have to be to accomplish the same purpose. In some cases, this additional complication is extreme, and an effort quite disproportionate to the simplicity of a basic algorithm goes into the design of specialized layouts of data structure and special schemes for reducing the number of passes over a tape which a calculation will require. Improved machines should progressively free programming from the burden of such cares. As programmers adjust their habits to improved equipment, their ability to cope with inherently complex situations should improve.

6.   Not only the physical computer but also the programming art has advanced during the past decade. An early but very basic concept is that of the *subroutine linkage,* whereby a generally useful subprocess — for example, taking a square root — could be separately programmed and routinely linked into a more comprehensive process. This technique makes it possible for an elaborate algorithm to be analyzed into simpler parts, which may then be programmed individually. Conversely, complex processes may rapidly be built up by the hierarchical combination of much more elementary logical parts. The utility of such a method is enhanced by its embodiment in a *programming source language.* Such a language is an assemblage of standardized elementary processes together with a grammar permitting the combination of these into composites. By standardizing programming technique, by providing for the automatic treatment of a host of detailed questions of data layout and access, by establishing succinct, mnemonic, and easily combined representations for frequently used processes, and by automatically checking programs for internal consistency and grammatical legality, such source languages make programming very much more effective than it would otherwise be.

The possibility of communicating with a computer through a programming language comes, of course, from the fact that the instructions a computer executes have the same internal format as the numbers with which it calculates. Thus, given the algorithm that describes the manner in which a particular programming language is to be expanded into a detailed set of internal instructions, a computer can calculate its own instructions from the indications present in a source language program. Moreover, since the phrases of a source language may be coded for internal representation within a computer, computers can also manipulate the statements of a programming language. Furnished with suitable algorithms, the computer is then able to accept definitions extending the grammar and vocabulary of a programming language and to retranslate extended statements into the basic statements they represent. Use of this technique permits an initial source language to be developed by successive extension for effective application to a given problem field. Searching for patterned repetitions in a given algorithm and embodying these patterns suitably in definitions may make it possible to express the algorithm efficiently and succinctly in a suitably extended programming language.

Another advantage of using programming languages for the communication of instructions to computers is that these languages are relatively machine-independent. Computers differing very greatly in their internal structure and repertoire of instructions can all be furnished with the same programs, each computer translating the source language into its own internal language. In particular, transition from a given computer to a later improved computer need require relatively little reworking of programs.

7. Experience has shown that various data structures (trees, lists, and so on) more complex than mere linear or multidimensional data arrays can be used effectively in the programming of elaborate algorithms. Various languages embodying useful general procedures for dealing with such structures have been constructed. Interesting methods have been developed for representing both complex data structures and algorithms for processing them with lists of a single universal form. Such methods preserve the complete flexibility of the underlying computer but allow its use at a considerably more organized level than that of internal machine language.

8. The continued development of source languages has made it possible to describe them to a computer largely in terms of the language itself. While the most basic elementary terms of any such description must be stated in a manner reflecting the specific instruction repertoire of a given computer, this "machine-dependent" portion of the description can be confined to a rather small part of the total information specifying the language. Thus, once the language is made available on a given computer, it can be carried over to a second computer by a relatively few redefinitions of basic terms, from which the first computer can proceed to regenerate the whole translation program in a form directly suitable for use on the second. If this "bootstrapping" technique, presently in a state of development and experimental use, could be suitably strengthened by the inclusion of program optimization algorithms, it might become possible for the translator to pass from a second computer to a third, and so on, and then back to the first without substantial loss of efficiency. Since each translation program carries along with it the body of programs written in the source language it translates, it would then become easier to assure continuity of development of the programming art.

The Computer as an Artificial Intelligence

The computer is not an adding machine but a universal information-processing engine. Its universality follows from the following basic observation of Turing: Any computer can be used to simulate any other. Thus, what one computer can do another can do also. Any computer is but a particular instance of the abstractly unique universal information processor. Computers differ not in their abstract capacities but merely in practical ways: in the size of the data base the computer can reach in a given time and in the speed with which a given calculation can be accomplished. Turing's observation applies, by virtue of its simplicity and generality, to any information-processing device or structure subject to the causal laws of physics, and thus to any information-processing device which is wholly natural rather than partly supernatural. It should in principle apply, therefore, to animal brains in general and to the human brain in particular.

From these reflections, one may derive a fascinating provisional goal for computer research: to duplicate all the capabilities of the human intelligence.

A crude inventory of the principal abilities that together constitute living intelligence will indicate the gross steps necessary in creating artificial intelligence.

1.   The human brain contains some 10 billion nerve cells, each capable of firing or failing to fire some hundreds of times per second. We may thus estimate that the information-producing capability of the brain is equivalent to the processing of approximately 1–10 trillion elementary bits of information per second. A computer of the largest current type contains some 500,000 elementary logical circuits, each of which may be used approximately 4 million times per second. Thus it processes approximately 2 trillion bits per second and, hence, is probably comparable to the brain in crude information-processing capability. Of course, the fact that individual neurons can perform rather more complex combinations of incoming signals than are performed by elementary computer circuits means that the foregoing estimates are uncertain, possibly by a factor of 100. Even if this is the case, however, computers available in the next few years should equal or exceed the crude information-processing capacity of the brain.

Access rates to memory are, however, much larger in the brain than in computers. Even assuming that the average neuron stores only 1 elementary bit of information, the brain has access to 10 billion bits, or 3 billion digits, of information at rates of access that do not impede its normal rate of function. Even a large computer of present design will not have access to more than 3 million digits of information at rates relatively as rapid. While much larger masses of information are, in fact, available to the computer, they are available only after substantial access delay.

Thus, while the ability of a large computer to process data probably equals that of the brain, its access to data is probably only 0.1 percent as satisfactory as that of the brain. These considerations reveal why computers perform excellently in iterative calculation with small amounts of data and in processing larger amounts of information used in simple sequences but are quite poor compared with the living brain in carrying out highly sophisticated processes in which large amounts of data must be flexibly and unpredictably combined. A considerable development of manufacturing technique will be required before it becomes feasible to furnish a computer with a rapid access memory more nearly comparable to that of the living brain.

A considerable development of the programming art also will be necessary before impressive artificial intelligences can be constructed. While it is true that much of the information used during the functioning of the adult brain is learned during earlier experience, it may also be surmised that a very large amount of genetic information contributes to the capacity to learn rapidly from fragmentary evidences. If, as a crude guess, we assume this inherent information comprises 10^{10} elementary bits, one for every neuron in the brain,[1] we may estimate the programming labor necessary to duplicate this information. It would be an excellent programmer who by present techniques is able to produce 300,000 correct instructions (10 million elementary bits of information) per year. Thus, one might estimate that to program an artificial intelligence, even assuming the general strategy to be clearer than it really is at present, is the labor of 1,000 man-years and possibly a hundred times more.

2. In the retina of the eye, the optic nerve, and the visual cortex of the brain the rapid and very extensive processing and reduction of the total pattern of light and dark sensed by the individual receptor cells of the eye take place. By a process currently understood only in bare outline, a vast mass of incoming visual data is reduced to a much smaller set of gestalt

---

[1]Note, for comparison, that a person with perfect recall, all of whose information was derived from reading at the rate of 200 pages per hour, 20 hours per day, 400 days per year, would have accumulated some 4×10^{12} elementary bits of information in a lifetime of 1000 years.

fragments or descriptor keys, which can be compared against an available stock of visual memories for recognition as a familiar object. A similar data reduction is performed in decoding auditory signals and is probably associated with the other senses as well.

The development of similarly sophisticated input programs for computers has been under way for some years and has attained modest success. It has become possible for computers equipped with optical scanners to read printed text with reasonable rapidity. A considerable effort in automatic spark-chamber and bubble-chamber photograph analysis has succeeded in developing programs capable of recognizing particle tracks and characteristic nodes in such photographs and of going on to analyze the geometry of perceived collisions. The evident economic importance of this visual pattern-recognition problem and the availability of faster computers better supplied with memory for experiments in pattern recognition assure continued progress in this area.

3. The brain coordinates the activity of elaborate assemblages of muscles in the apparatus of speech and in the body at large. This effector control is combined with sophisticated analysis of each momentary pattern of autosensation, permitting the stable and effective performance of tasks in a varying environment. The absence until two decades ago of mechanical brains of any kind other than crude mechanical clockworks and rudimentary electric feedback circuits has unfortunately caused mechanical design to be concerned more with mechanisms for the rapid repetition of ultrastereotyped tasks than with the development of highly flexible or universal mechanisms. The coordinating "mind" computers furnish thus finds no really suitable "body" ready to receive it. Nevertheless, the increasing use of computers in industrial and military process control and the association of computers with various mechanical and electric effectors in "real-time" scientific experimentation have begun to extend our experience with the problem of attaching muscles to the computer brain. A few interesting special research efforts in this area are also under way.

4. A number of interesting experiments have shown that computers can learn if they are programmed with criteria of success in performance and with ways to use these criteria to modify the parameters determining behavior choice in situations too complex for exhaustive calculations to be feasible. Such a technique has been used successfully to train character-reading programs to discriminate between letters printed in a given type face and to allow a checker-playing program to improve its ability by studying past games or by practicing against a copy of itself. The learned adaptations in these cases are, however, modifications of parameter in a program of fixed structure rather than acquisitions of more fundamental behavior patterns. An important basic method in natural learning is the discovery of significant similarities in situations that are not nearly identical. This fundamental activity of the mind has not been duplicated on the computer in any striking way. We do not yet have methods whereby the vitally important process of learning basic general principles from fragmentary indications can be made to function effectively.

5. A structure of motive and emotion, ultimately serving the purpose of individual and group preservation, necessarily forms a part of the living brain. Motives of preservation and predation have perhaps been the principal forces in the evolution of intelligence. (Plants, not being predators, have never developed intelligence.) Nevertheless, it is the ability to perform a set task successfully and rapidly, not the will either to perform it or not to perform it, which measures intelligence. At any rate, as computers grow in ability and steadily come to occupy more strategic positions in economic and social function, it becomes more desirable that they have no programmed motive other than that of advancing the general purposes of the institution they serve.

6. The elements of intelligence mentioned are all part of the inheritance we share with other mammals and even with animals generally. Certain additional aspects of intelligence, while existing in relatively primitive forms in animals other than man, are much more highly developed in man. These may

perhaps be typified by the understanding of language and the ability to deal with abstract concepts.

A great deal of work has gone into developing the linguistic abilities of computers. Much in now understood about how underlying grammars permit the reconstruction of relationships implicit in a sequence of words or symbols. Such understanding is currently applied in a routine way to the interpretation of mechanical "source languages" with narrowly limited vocabularies and somewhat restricted grammars. Languages of this kind have become the standard vehicle of man-computer communication. Work on natural languages, with their large vocabularies and considerably more elaborate grammars, has also gone forward with fair success, constrained, however, by the complexity of the resulting programs and by the limitations of computer memory. Such work, incidentally, would benefit greatly from any substantial progress in the programming of computer learning, which would make it possible for computers to acquire their grammar, as people do, by studying source text. There is, all in all, no reason to anticipate fundamental difficulties in developing the computer's ability to understand language as far as resources allow.

Arithmetic and Boolean logic form the computer's basic reflexes, so that there is a real sense in which the computer's ability to deal with abstract concepts is inherent. Experiments in theorem proving by computer may be considered as explorations of this issue in a context of reasonable complexity. These studies have succeeded in devising methods whereby the computer can manipulate general concepts, discover the proofs of general statements, and even guide its conceptual explorations by heuristic principles of elegance and effectiveness. Nevertheless, they flounder relatively quickly because of the computer's present inability to sense useful similarities in sufficient breadth. This, rather than any difficulties associated with abstractness *per se,* seems to be the principal block to improved performance. If a computer could find worms on a twig with the effectiveness of a bird, it might not fall so far short of the mathematician's ability to hunt out interesting theorems and definitions.

Prospects at Long Term

Humanity is constrained, not in its tools but in itself, by the infinitesimal pace of evolution. If we are able to create a human intelligence, we shall shortly thereafter be able to create highly superhuman intelligences. These may have the form of huge self-running factories producing varied objects in response to a stream of incoming orders; or of groups of roving sensor-effectors, gathering external visual or other data, reducing them for transmission and subsequent analysis to a central brain, and acting in accordance with retransmitted instructions, the whole forming a kind of colony or army of telepathic ants; or of an electronic mathematician, selecting interesting consequences from the association of imaginations at a superhuman rate.

It is apparent that such devices would have an enormously profound effect on all the circumstances of human society. It seems to the author that present computers are, in fact, the earliest species of a new life form, which may be called *crystallozoa* in distinction to the older *protozoa*. In the near future, on a historical scale, the crystallozoa will be able to store information with the same molecular-level density as do the protozoa. They will then enjoy advantages corresponding to the 10^6:1 ratio of electronic/protoplasmic clock rates or to the 10^4:1 ratio between the speed of light and the speed of signal transmission along neurons. If any of this is true, it is clear that what has been seen until now is but the faintest indication of what is to come.

1969

References

1. MARVIN MINSKY, *Computers: Finite and Infinite Machines,* Prentice-Hall, Englewood Cliffs, N.J., 1967.
2. MARTIN DAVIS, *Computability and Unsolvability,* McGraw-Hill, New York, 1958.
3. E. FEIGENBAUM AND J. FELDMAN, *Computers and Thought,* McGraw-Hill, New York, 1963.
4. DEAN E. WOOLDRIDGE, *The Machinery of the Brain,* McGraw-Hill, New York, 1963.
5. DAVID H. HUBEL AND T. N. WIESEL, "Receptive Fields of Single Neurons in the Cat's Striate Cortex," *Journal of Physiology,* 148, 574-591 (1959).

CHAPTER EIGHT

Mathematics: Trends

Introduction

UNLIKE DISCIPLINES WITH EMPIRICAL BACKGROUNDS, mathematics lacks central problems that are clearly defined and universally agreed upon. As a result, the development of mathematics proceeds along a number of seemingly unrelated fronts, which tends to present a picture of fragmentation and division. Adding to the difficulty of evaluating its present state and of (guardedly!) predicting its future, is the fact that during the past few decades mathematics became increasingly isolated from its sister disciplines, and as a result of turning inward there was a marked increase in the level of abstraction and a reinforcement of the ever-present trend to greater and greater generality.

Ironically, the isolationist and introspective trend in mathematics coincided with a veritable explosion in demand for its "services" to an ever-widening circle of "users." There is now hardly a corner of human activity which has not for better or for worse been affected by mathematics. So much so, in fact, that one may speak with considerable justification of a "mathematization" of culture.

Ever since Plato's invectives against Eudoxus and Archytas there have been tensions between the two main trends of mathematical creativity: the abstract ("unembodied objects of pure intelligence") and the concrete ("means of sustaining experimentally, to the satisfaction of the senses, conclusions too intricate for proof by words or diagrams"[1]). The tensions between the two trends (more easily recognized under the names of *pure* and *applied*) persist to this day, aggravated by the emerging power of the computer and complicated by socio-economic factors. Thus Professor Marshall Stone, writing in 1961 asserted that "while several important changes have taken place since 1900 in our conception of mathematics or in our points of view concerning it, the one which truly involves a revolution in ideas is the discovery that mathematics is entirely independent of the physical word" [1]. A few lines later Professor Stone elaborates:

> When we stop to compare the mathematics of today with mathematics as it was at the close of the nineteenth century we may well be amazed to see how rapidly our mathematical knowledge has grown in quantity and in complexity, but we should also not fail to observe how closely this development has been involoved with an emphaisis upon abstraction and an increasing concern with the perception and analysis of broad mathematical patterns. Indeed, upon close examination we see that this new orientation, made possible only by the divorce of mathematics from its applications, has been the true source of its tremendous vitality and growth during the present century.

No one will deny that mathematics can reach ultimate heights in splendid isolation. One does not even have to look to our century for striking examples. Surely Galois theory, culminating in the proof that algebraic equations of degree 5 and higher are, in general, not solvable in terms of radicals, is gloriously convincing.

There are even examples, again taken from the past, of

---

[1]I am referring here to the following passage in Plutarch's *Life of Marcellus:* "Eudoxus and Archytas had been the first originators of this far-famed and highly prized art of mechanics, which they employed as an elegant illustration of geometrical truths, and as means of sustaining experimentally, to the satisfaction of the senses, conclusions too intricate for proof by words and diagrams... But what of Plato's indignation at it, and his invectives against it as mere corruption and annihilation of the one good in geometry, "which was thus shamefully turning its back upon the unembodied objects of pure intelligence to recur to sensation and to ask help (not to be obtained without base supervisions and depravation) from matter..."

mathematical developments which, having begun with a concern for "unembodied objects of pure intelligence," have
ultimately become instruments of deeper understanding of the
physical universe. I have in mind especially the General Theory
of Relativity, which in its mathematical apparatus depends
crucially on differential geometry, a branch of mathematics
which owes its inception and much of its early growth to non-
Euclidean geometry. There is, however, an implication in Professor Stone's thesis that isolation is also *necessary* for survival
and growth of mathematics, and this I find totally unacceptable.

On a number of occasions, I made reference to a wartime
cartoon depicting two chemists surveying ruefully a small pile
of sand amidst complicated looking pieces of equipment. The
caption read: "Nobody really wanted a dehydrated elephant,
but it is nice to see what can be done." Forced isolation of
mathematics, though it may result in many spectacular
achievements, will also proliferate dehydrated elephants to a
point where the price may be too high to pay.

With all this as background, I shall attempt to give a glimpse
of some of the prevailing themes, trends, and tensions in contemporary mathematics. The picture will be fragmentary and
incomplete, but this I am afraid is unavoidable. The limitations
of one's competence and the very nature of modern mathematics
are too formidable to overcome.

Scaling the Heights

In an article "American mathematics from 1940 to the day before
yesterday" [2] six authors selected (according to certain clearly
stated rules and with due apologies for biases and omissions)
ten theorems which they considered to have been pinnacles of
achievement of American mathematics in the period from 1940
to 24 January 1976, i.e. the day Professor Halmos delivered an
abbreviated version of the paper as an invited address before
a meeting of the Mathematical Association of America. Since
much of the material depended significantly on the work of non-
Americans (underscoring the truly international character of
mathematics) it is safe to say that the selection can be taken

79

as representative of the achievements of mathematics the world over. The authors by one of their rules included only "pure" mathematics, and there is little doubt that their picture of this part of mathematical landscape is photographically accurate.

All the theorems confirm, with vengeance I may add, that "mathematics is entirely independent of the physical world," thus lending support to Professor Stone's neo-Platonist view of advantages of isolation. On the other hand, the very fact that it required a collaboration of *six* very able and highly accomplished mathematicians to merely explain to their fellow practitioners the backgrounds and formulations of the ten most spectacular achievements of their subject is in itself a telling commentary on the state of the art. Of the ten, I have chosen three for a brief presentation. My choice is dictated by the belief that these are less "internal" to mathematics than the others and hence perhaps of greater general interest.

The first of the three is the independence of the continuum hypothesis from the standard (Zermelo-Fraenkel) axioms of set theory, and it represents a solution to the first on the famed list of unsolved problems posed by Hilbert in 1900. The question goes back to Georg Cantor, the creator of set theory. Cantor defined two sets to be of equal cardinality ("being equally numerous") if a one-to-one correspondence between the elements of the two sets could be established. A set A is said to be of greater cardinality than a set B if there is a *subset* of A of equal cardinality with B, while the cardinality of A itself is not equal to that of B. In other words, there is a subset of A whose elements can be put into one-to-one correspondence with those of B, but there is no such one-to-one correspondence between all the elements of A with those of B.

For finite sets, equal cardinality simply means that the sets contain the same number of elements. For infinite sets, the situation is more subtle since a subset of an infinite set can have the same cardinality as the set itself. For example, the set of even integers, though clearly a subset of the set of all integers, is of the same cardinality as the larger set, since a one-to-one correspondence between the elements of the two can be established as:

$$
\begin{array}{ccccc}
2 & 4 & 6 & 8 & 10 \quad \cdots \\
\updownarrow & \updownarrow & \updownarrow & \updownarrow & \updownarrow \\
1 & 2 & 3 & 4 & 5 \quad \cdots
\end{array}
$$

The set of (positive) integers is the "smallest" infinite set, i.e. its only subsets of smaller cardinality are finite. All sets of the same cardinality as the set of integers are called denumerable. It was shown by Cantor that all rational numbers (i.e. numbers of the form p/q, with p and q integers without a common divisor) form a denumerable set. Much more surprisingly, Cantor has also shown that the set of all real numbers is not denumerable and hence of greater cardinality than that of the set of integers. The proof is simple and is so well known that I reproduce it only with profound apologies to the overwhelming majority of my readers. Since I want to comment on it later on, I thought it best to have it in front of one's eyes rather than rely upon one's memory.

Every real number can be uniquely written as a non-terminating decimal (this requires, for example, writing $6/5 = 12/10 = 1.2$ in the equivalent way as $1.1999\ldots$). If real numbers could be written as a sequence (which is another way of saying that if the set of real numbers were denumerable), we could have a tableau

$$
\begin{array}{l}
c_{11} \; . \; c_{12} \; c_{13} \; c_{14} \cdots \\
c_{21} \; . \; c_{22} \; c_{23} \; c_{24} \cdots \\
c_{31} \; . \; c_{32} \; c_{33} \; c_{34} \cdots \\
\qquad \cdot \quad \cdot \quad \cdot \quad \cdot \\
\qquad \cdot \quad \cdot \quad \cdot \quad \cdot \\
\qquad \cdot \quad \cdot \quad \cdot \quad \cdot
\end{array}
$$

containing all of them. But the number

$$
d_1 \; . \; d_2 \; d_3, \ldots \;,
$$

where $d_1 \neq c_1$, $d_2 \neq c_{22}$, $d_3 \neq c_{33}$, etc. is clearly different from all elements of the tableau and thus we have a contradiction which means that the set of all real numbers is of *greater* cardinality than the set of integers.

The continuum hypothesis of Cantor is the assertion that the set of real numbers is the *next* largest after the set of integers or, in other words, there are no sets whose cardinality is greater than that of the set of integers and less than that of the set of real numbers.

There is a historical analogy between the fate of the continuum hypothesis and Euclid's fifth postulate. Cantor believed his hypothesis to be nearly self-evident. As doubt arose, an enormous amount of work was undertaken in constructing statements logically equivalent to Cantor's hypothesis or its negation in the search for one so bizarre that an acceptance or rejection of the hypothesis could be forced. How very much like attempts of Saccheri and his fellow toilers in their search for a self-evidently rejectable consequence of a negation of the fifth postulate!

In both cases the labor was in vain. For, as Paul J. Cohen showed in 1963, paralleling the feat of Bolyai and Lobachevsky, one can keep or reject the continuum hypothesis without any fear of logical contradiction. Here the analogy ends. For as non-Euclidean geometries became part of the mainstream of mathematics and lived to influence profoundly our perception of the physical world, the non-Cantorian set theories have not, so far at least, had any discernable effect on the development of mathematics, let alone on our views of the physical universe.

Is it possible that the strange non-Cantorian creations could become frameworks for some future physical theories? Personally I doubt that this will ever be the case. My main reason for feeling doubtful is that the independence of the continuum hypothesis can be restated as an assertion that there are models (realizations) of the set of real numbers of arbitrarily high cardinality. In other words, the usual collection of axioms which describe the properties of real numbers are logically consistent with the *set* of real numbers being of arbitrarily high cardinality (higher, of course, than the cardinality of integers), or in yet other words, there are non-Cantorian models of the set of reals.

The set of reals has been a troublesome concept from the

beginning. Cantor's striking proof that it is not denumberable led immediately to a paradoxical conclusion. Since the set of numbers which can be defined in a finite number of words can be easily shown to be denumerable, it followed that there must be real numbers which *cannot* be defined in a finite number of words. Can one sensibly speak of objects definable in a sentence of infinite length? The great Henri Poincaré thought that the answer should be an emphatic one. *"N'envisagez jamais que des objets susceptibles d'être définis dans un nombre fini de mots"* he proclaimed, and there is a whole school of logic (the intuitionists) which lives by this dictum. So it all boils down to whether we have found a proper axiomatic description of reals, and this question appears closer to philosophy than to physics. Nevertheless, one cannot tell and there are serious mathematicians (S. M. Ulam is one) who would disagree with me.

One final remark. Cantor's diagonal construction, which is at the heart of his proof that the set of reals is non-denumerable, has to be treated with care, for it can be turned into a source of paradoxes. The following (a variant of a paradox of Jules Richard) is a good illustration.

We say that a function f defined for all integers 1, 2, 3, ... and assuming only integer values is *computable* if for every integer n there is a *prescription* consisting of a finite number of words which allows one to calculate $f(n)$ in a finite number of steps. One can easily show that the computable functions form a denumerable set, and they can thus be arranged into a sequence

$$f_1, f_2, f_3, \ldots.$$

Consider now the function g defined by the formula

$$g(n) = f_n(n) + 1$$

Now $g(n)$ is clearly computable and yet since $g(1) \neq f_1(1)$, $g(2) \neq f_2(2)$, etc. it is not included in our sequence: a contradiction!

The source of the paradox is subtle, and it has to do with the fact that we are asking whether g is a member of a set in terms of which g itself is defined. That such circularity is fraught with danger has been known since antiquity, certainly since

Epimenides the Cretan proclaimed that all Cretans lie.

Let me now turn to my second illustration. It again involves the solution of a problem of long standing and one which also appeared (as No. 10) on Hilbert's list of 1900. The problem concerns diophantine equations, i.e. polynomial equations with integer coefficients for which we seek only integer solutions. The simplest is

$$ax + by = 1,$$

(*a*, *b*, *x*, and *y* being positive or negative integers) and its solution was already known to the Greeks. The next simplest is the so-called Pell's equation

$$x^2 - Dy^2 = 1,$$

where *D* is a positive integer. The complete theory of this equation has been known for over two hundred years.

In both of these cases there are simple algorithms for generating all integer solutions. In the linear case it is related to the so-called Euclid algorithm; in the quadratic case the algorithm is to expand $\sqrt{D}$ in a continued fraction:

$$\sqrt{D} = a_0 + \cfrac{1}{a_1 + \cfrac{1}{a_2 + \dots}}$$

Consider the continued fraction

$$\frac{P_n}{q_n} = a_0 + \cfrac{1}{a_1 + \cfrac{1}{a_2 + \cfrac{\dots}{\cfrac{1}{a_n}}}}$$

A solution to Pell's equation can then be obtained simply from the $p_n s$ and $q_n s$.

What is an algorithm? Without entering into technicalities it is sufficient to identify algorithms with computer programs. It is quite easy to program a computer to generate solutions of the linear and quadratic diophantine equations, but for equations of higher degrees and involving more than two variables the situation is clearly much more complicated.

Hilbert's tenth problem was to prove (or disprove) that, given a (polynomial) diophantine equation, an algorithm can be found to generate all of its solutions. The answer is in the negative, i.e. there are diophantine equations for which no algorithm to generate all of its solutions can be found.

The theorem is purely existential, i.e. it merely asserts that certain objects *exist* without providing us with a *concrete* prescription of how to exhibit them. In fact, the Cantor diagonal construction is used in the final stage of the proof, thus making it non-constructive in the usual sense of the word.

"Impractical" as the negative solution of Hilbert's tenth problem is, it was not without influence on the rapidly growing discipline of computer science. It brought the power of modern mathematical logic to bear upon questions of computability, and it brought these seemingly mundane questions closer to foundations of mathematics and even to philosophy.

So far my illustrations were from the realm of foundations. My third (and last) is from the heart of the mathematical superstructure. The actual theorem chosen by Halmos *et al.* is the so-called index theorem of Atiyah and Singer, which is rather technical. It has to do with certain elliptic (differential) operators defined on manifolds, and it culminates in proving that an integer defined in terms of topological properties of the manifold is equal to another integer defined in terms of the operator. Although the motivation for the Atiyah-Singer index theorem came from algebraic geometry (the theory of Riemann surfaces to be more specific) the theorem is one of a class of theorems connecting differential equations and topology, whose roots can also be traced to physics. More importantly, the physical background can be turned into a potent mathematical weapon, thus reminding us of the words of Poincaré: *"La Physique ne nous donne seulement l'occasion de résoudre des problèmes..., elle nous fait présenter la solution."*

Let me very sketchily and in a greatly oversimplified manner try to describe what is involved. In ordinary Euclidean space

— let us say the plane — heat conduction is described in appropriate units by the differential equation

$$\frac{\partial P}{\partial t} = \Delta P = \text{div (grad } P) = \frac{\partial^2 P}{\partial x^2} + \frac{\partial^2 P}{\partial y^2} \, ,$$

where $P(x, y, t)$ is the temperature at (x, y) at time t.

If we consider a bounded region whose boundary is kept at temperature 0, then starting with an arbitrary temperature distribution $P(x, y; 0)$ at $t = 0$ we shall observe an approach to 0 of the form

$$P(x, y; t) = \sum_{n=1}^{\infty} c_n(x,y)\exp(-\lambda_n t) \, ,$$

where the $\lambda_n s$ are positive numbers. The terms $\lambda_1, \lambda_2, \ldots$ are called the eigenvalues of the Laplacian Δ, and it turns out that

$$\Delta c_n + \lambda_n c_n = 0,$$

with $c_n = 0$ on the boundary of the region. (If the region were a membrane held fixed along its boundary, the λ_n would be the squares of frequencies of fundamental tones which the membrane was capable of producing.)

Let us now consider the same heat-conduction problem but on a smooth, closed, two-dimensional surface (e.g. the surface of a sphere of radius R).

$$\frac{\partial P}{\partial t} = \text{div(grad } P) = \Delta P \, ,$$

but Δ is now called the Laplace-Beltram operator (it can be expressed explicitly in terms of curvilinear coordinates) and there is no boundary, so that $\lambda_0 = 0$ and an arbitrary temperature distribution will approach the uniform distribution on the surface.

It is a natural question to ask what information about the surface can be obtained from the $\lambda_n s$. The answer, striking in its simplicity and beauty, is contained in the following formula of McKean and Singer [3] valid for small t:

$$\sum_{n=0}^{\infty} \exp^{(-\lambda_n t)} = \frac{S}{4\pi t} + \frac{1}{6} E + \text{terms which approach 0 as } t \to 0 \, .$$

Here S is the area of our surface and E the so-called Euler-Poincaré characteristic, a topological invariant which can be defined in an elementary way as follows:

Consider a polyhedron inscribed in the surface and let F, E, and V be respectively the number of faces, edges, and vertices of this polyhedron. The Euler-Poincaré characteristic is then given by the formula

$$E = F - E + V.$$

It does not in fact matter that the polyhedron is inscribed; it suffices that it be *continuously deformable* into the surface. For a sphere $E = 2$ and for a torus $E = 0$.

The theorem connecting the eigenvalues of the Laplacian and the Euler-Poincaré characteristic has been generalized (to manifolds of arbitrary even dimension) and extended in many directions.

Perhaps if one compares the circles of ideas involved in this problem and in the Cantor problem discussed at the beginning of this section one might gain a measure of appreciation of the unbelievable breadth of mathematical problems. One might also wonder as to how the existence of non-Cantorian systems could in any way affect the theory of heat conduction or the eigenvalues of the Laplacian.

Finding Common Ground

I have already mentioned that during the past few decades mathematics has largely developed in isolation. Whether as a result of a revolution, as Professor Stone would have us believe, or simply because internally generated problems provided a sufficiently exciting intellectual fare, the fact remains that mathematics all but severed its ties with sister disciplines.

Even in the past much of the development of mathematics was mostly internal and wholly autonomous. But the lines of communication between mathematics on the one hand, and physics and astronomy on the other were open, so that a healthy flow of ideas in both directions could be maintained.

Thus when in about 1914 Einstein struggled with formulating field equations of general relativity, he soon found out that a mathematical apparatus that he needed was already in existence in the form of so-called Ricci calculus, an outgrowth of Riemann's inspired *Habilitationsvortrag "Über die Hypothesen welche der Geometrie zugrunde liegen."* Similarly when the new quantum mechanics was being created in the 1920s and 1930s, some of the necessary conceptual and technical mathematical background was again available in the form of the theory of Hilbert spaces, although much of the theory had to be developed further to fit the needs of the new physics.

The Ricci calculus, which is a branch of differential geometry, is directly traceable to non-Euclidean geometries. Its development was internally motivated and constituted a natural generalization and extension of the work of Gauss and Riemann. Riemann expected that his geometry could some day intervene in problems of physics and astronomy, but neither he nor his followers could have imagined the spectacular and dramatic way in which it ultimately did. The theory of Hilbert spaces on the other hand, evolved from problems of nineteenth-century physics and culminated in the miracle of unifying atomic spectra and fundamental tones of membranes and strings.

The interweaving of mathematical and physical ideas in general relativity and in quantum mechanics belongs to the most glorious chapter of the history of natural sciences. It is therefore truly surprising that the partners in these great adventures of human mind gradually came to a parting of ways.

It is not that mathematics abandoned either differential geometry or the theory of Hilbert spaces. Quite to the contrary, both continued to thrive. It is just that the preoccupations, the problematics, and the directions of mathematics and physics diverged to such an extent that the resulting separation seemed irreversible. Fortunately we may be witnessing a change in the trend and, though it is too early to tell, we may in fact be in the midst of a period in which ideas which have originated in vastly disparate contexts are beginning to show signs of yet another miraculous confluence. Interestingly enough, the mathematical ideas have come from a far-reaching extension of those underlying Riemannian geometry, itself an inspired extension of the theory of curved surfaces. It is difficult without considerable technical preparation to give an honest account of what is really involved, but I will try to give the reader a little taste of the brew.

First though, a quick glance at the theory of surfaces, illustrated on the example of the surface on a sphere of radius R in ordinary three-dimensional Euclidean space.

Introducing polar coordinates (with ϕ the latitude and θ the longitude) we have

$$x = R \sin \phi \cos \theta;$$

$$y = R \sin \phi \sin \theta;$$

$$z = R \cos \phi .$$

The Euclidean distance ds between two infinitesimal close points is (from Pythagoras' theorem)

$$ds^2 = dx^2 + dy^2 + dz^2 .$$

If the points are constrained to lie on the surface of our sphere, one has

$$ds^2 = R^2(d\phi^2 + \sin^2\phi \, d\theta^2) .$$

At each point P of the surface of our sphere we consider the plane tangent to it, and a vector $\vec{v}(P)$ in that plane which changes smoothly as P moves among the surface. How do we define the differential of $\vec{v}(P)$ as P changes infinitesimally? One would be inclined to merely take the difference $\vec{v}(P') - \vec{v}(P)$ (P' and P infinitesimally close). But this difference would also have a component in the direction perpendicular (normal) to the surface of the sphere, and one takes the view that a flat creature confined to live on that surface would be incapable of detecting changes in the normal direction. We thus define the differential D as the tangential component of the usual differential d, and call it the *covariant differential*. This covariant differential leads to the *covariant derivative* by the formula

$$\frac{D\vec{v}}{ds} ,$$

ds being the differential distance between P and P'.

If P moves along a curve in such a way that the covariant derivative of $\vec{v}$ is zero, we say that $\vec{v}$ is undergoing *parallel displacement*. If $\vec{v}(P)$ is parallel displaced along a closed curve it may not come back to its original value. In such a case the surface is curved, and one can define (and describe) its curvature in terms of the changes suffered in parallel displacement.

All this (in a different terminology) was already known to Gauss. Gauss made one more observation, which he recognized as being of crucial importance. (The observation is embodied in a theorem he called Theorema Egregium.) The observation was that the concept of curvature is *intrinsic*, i.e. it can be defined without any reference to the Euclidean space in which the surface is imbedded. In other words, the hypothetical flat creature could define curvature without knowing that there is a world in the direction normal to the surface on which it lives.

Riemann took the next step by replacing surfaces by more general manifolds, which could be coordinatized (at least locally) and endowed with a metric

$$ds^2 = \Sigma g_{ij} dx^i dx^j$$

(by analogy with such formulae for surfaces) which *need not* have been "inherited" from an Euclidean metric in the way

$$ds^2 = R^2(d\phi^2 + \sin^2\phi d\theta^2)$$

was "inherited" from

$$ds^2 = dx^2 + dy^2 + dz^2 .$$

The Italian geometers developed this idea further by extending the concepts of covariant differentiation, parallel displacement, and curvature. Unlike in the case of surfaces in Euclidean three space, in which curvature could be described by scalars, the higher-dimensional cases required tensors for a full description of how the space is curved. By the time Einstein needed a theory of curved spaces, it was there for him to use.[2] Riemannian geometry, rooted as it was in the theory of curved surfaces, made essential use of tangent spaces and defined covariant differentiation and parallel displacement by analogy with that theory. The underlying skeleton is the space (manifold) M with a space T_x(tangent space) attached to every point x of M and a rule (connection) which tells us how to perform parallel translations on elements of T_x as x moves in M.

I have just given a vague and imperfect description of a fiber bundle which became the stepping-stone to the next development of differential geometry. Tangent spaces (fibers) can be replaced by more complicated structures not necessarily related to the space to which they are attached. Similarly, connections need not be defined in terms of metric properties of the underlying space. To be sure some restrictions have to be imposed, and these are dictated by a degree of adherence to the parental Riemannian geometry.

But the newly gained freedom is great — so great in fact, that the familiar theory of electromagnetism can be couched in geometric terms. Vector potential intervenes in the definition of covariant differentiation and the electromagnetic field tensor becomes the curvature tensor in the new formulation. The much studied Yang-Mills fields also fit neatly into the new scheme, and results of direct physical relevance are uncovered in the intricate web of abstract mathematical speculations.

To whom does the computer belong?

The usefulness, even indispensability, of the computer as a tool

---

[2]Riemannian geometry was developed for the case of *definite metric*, i.e. the case in which $\Sigma g_{ij} dx^i dx^j$ is positive definite (in other words for locally Euclidean spaces). In relativity theory the space (or rather space-time) is locally Minkowskian, i.e. by a suitable change of local coordinates $\Sigma g_{ij} dx^i dx^j$ can be transformed into

$$(dy^1)^2 - (dy^2)^2 - (dy^3)^2 - (dy^4)^2 .$$

of science and technology is now universally recognized and acknowledged. Imaginative uses of this remarkable instrument keep multiplying daily, covering an ever-widening spectrum of fields and disciplines. Even the humanities have not escaped — as witness concordances on which studies of stylistic peculiarities can be based. These can now be compiled and collated in a matter of weeks or even days, whereas they formerly required years of tedious and dull labor.

In a sense, the computer belongs to all of its users — even those who do not use it wisely or honestly. But the deeper question is: where does the computer belong in an intellectual sense? To put the question differently, what are the questions of general intellectual (as opposed to merely utilitarian) interest generated by the existence of the computer and to what, if any, discipline should they be assigned?

The question is not easy to answer because on the one hand the important (and certainly challenging) problems concerned with programming languages and related questions of "software" do not easily fit into existing disciplines (if anywhere they belong somewhere between linguistics and logic), while on the other hand problems centering on the concept of algorithmic complexity, which unquestionably belong to mathematics, could have been formulated before computers, as we know them today, came into existence.

Since my concern is mainly with mathematics, let me say a few words about algorithmic complexity, especially since problems concerning this concept are, I believe, the only problems formulated in this century which are not traceable to the last one. It will facilitate discussion if I begin by stripping the computer of all its electronic gear and programming complexities, and return to its ancestor, namely the universal Türing machine.

The Türing machine is an infinite tape, subdivided into equal adjacent squares, each square being either blank (denoted by *) or having a vertical bar (|) in it. There is also a scanning square which can be moved to the right (r) or left (l), one step (i.e. square) at a time. One can also replace (R) the symbol in the square by the other one, i.e. * by | and vice versa. Finally (and all-importantly) one can halt (h) the whole procedure.

A program is a (finite) set of numbered instructions of the form

$$12 : * R \ 8 \ ,$$

which is instruction 12 and reads: if the scanned square is blank print a vertical bar and look up instruction 8. This is all!

It has been proved that every algorithm (which is defined in terms of so-called recursive functions) can be programed on a Türing machine. The minimal length of a program is then the measure of the complexity of an algorithm. Thus we have a way of ranking algorithms by their complexities. Similarly we can define the complexity of a task as the minimum number of steps it takes to perform it on a Türing machine. One can then ask, for example, whether adding two n-digit numbers is of comparable complexity as multiplying such numbers — a question that is difficult and I believe still unanswered.

The subtle and elusive concept of randomness has also been coupled with that of complexity. A. N. Kolmogorov and, independently, Gregory J. Chaitin have proposed that a sequence of zeros and ones of length n be defined as "random" if its complexity (i.e. the length of the shortest program needed to generate it) is of order n. This definition is tenable only in an asymptotic sense, i.e. in the limit as n approaches infinity, and is beset by a number of interesting logical difficulties [8]. Still, the idea is an interesting one, and for a while it received considerable attention.

Estimating complexities of algorithms or tasks is extremely difficult, and few general methods are available. The problem is also of considerable practical importance. Has the computer, however, been instrumental in solving an important purely mathematical problem?

The four-color problem comes to mind, but here the major step in the solution was the reduction of the problem of coloring (with four colors) of *all* maps to coloring a *finite* number of specific ones. The latter task was relegated to the computer which came through with flying colors, completing the tedious and relatively uninteresting part of the proof.

Adherents of "artificial intelligence" notwithstanding, it is difficult to see how the computer can be genuinely creative, although I admit that my scepticism may be vacuous, since nobody knows what "genuinely creative" really means.

Computer experimentation may on the other hand be very illuminating and lead the mathematician to new and interesting fields to explore. Recently I found a particularly convincing illustration. It starts from a problem suggested by ecology, and what follows is based on two papers by R. M. May [9,10].

For certain species population growth takes place at discrete time-intervals, and generations are therefore non-overlapping. Assuming that the population N_{t+1} during the time-interval $(t + 1) - st$ depends only on the population N_t during the preceding time-interval, we are led to a difference equation

$$N_{t+1} = F(N_t) ,$$

where the function F is chosen to conform with biological intuition and knowledge.

The simplest case is the linear one

$$N_{t+1} = \exp(r)N_t ,$$

where r is the growth rate. The solution is

$$N_t = \exp(tr)N_0 ,$$

which for $r > 0$ leads to the Malthusian explosion and for $r < 0$ to extinction. If the environment has carrying capacity k (i.e. as soon as the population exceeds k it must decrease), the linear model must be abandoned, and the following simple non-linear model is one of a number that have been proposed

$$N_{t+1} = N_t \exp \left[r \left(1 - \frac{N_t}{k} \right) \right] .$$

There is now no simple formula for N_t and the situation is already extremely complicated. If $0 < r < 2$, N_t approaches as t approaches infinity the carrying capacity, k, no matter what the initial population N_0 is. As soon as r exceeds 2, the population size N_t (again for t approaching infinity) shows a marked change in behavior, for it now oscillates between two numbers exhibiting what is called a two-point cycle. Then comes the next critical value of r, namely $2 \cdot 656$ ... above which a four-point cycle is observed, then above $2 \cdot 685$ ... an eight-point cycle, etc. The critical numbers seem to approach $2 \cdot 692$ ..., and as soon as r exceeds this value, N_t fails to exhibit any kind of regularity and becomes reminiscent of sample traces of discrete stochastic processes. This "chaotic" regime is not yet very well understood, but the phenomenon is striking and it poses problems of great mathematical interest. Much progress has however been made in providing a rigorous mathematical theory of the prechaotic behavior, but even here the full story is far from finished.

Although I firmly believe that the computer's intellectual home is in mathematics, it is not a belief that is now widely shared within the mathematical community. One detects though, here and there, signs of awakening tolerance, and one can only hope that success stories like that of population-growth models will contribute to full acceptance of the computer as a member of the family. Mathematics could use a healthy dose of experimentation, be it only of the computer variety.

Where from and where to

I have already stated in the Introduction that to give an honest and even moderately complete account of the present state of mathematics is an impossible task. So much is happening on all fronts, so much of it is internal to this or that field or subfield, and so much is a reflection of passing vogues that even if a review of acceptable accuracy were possible, it would be likely out of date in a year's time. Still, the enormous variety of preoccupations, which by some unstatable criteria are classed as mathematics, is holding together.

Is it possible to single out major developments or trends (as opposed to conquests of problems of long standing) which have had more than a local impact on this or that branch of mathematics?

Fifty, sixty years ago the theory of linear spaces would have been considered as such a development. This was the great period of geometrization of mathematics, and the aftershocks are still with us. Then came a period of algebraization of mathematics, and unification was sought by fitting as much mathematics as possible into algebraic structures: groups, rings, ideals, categories, etc. This trend is still very much with us, and the algebraic appearance of much of today's mathematics is perhaps its most notable characteristic.

This in a sense is an *organizational* trend — a trend to organize mathematics along formal algebraic lines. It has certainly been successful up till now, and it will certainly leave its mark on the mathematics of the future. And yet there is in such organizational successes an element of Monsieur Jourdain's surprise that one has been speaking prose all one's life.

Except for such generalities it is difficult to detect all encompassing trends. No doubt it is because an overall unifying direction is missing that one is sensitized to confluences of ideas coming from diverse sources, especially if one of the sources happens to lie outside of mathematics. Thus the excitement about theorems like the index theorem (linking geometry and differential equations) or the discovery of topological aspects of gauge fields, which brought fiber bundles to the peripheries of physics.

Let me deal briefly with yet another confluence, confessing at the outset to a strong personal prejudice. I have in mind the gradual infusion of probabilistic ideas and techniques into different parts of mathematics.

Probability theory had an unusual history. After a rather inauspicious beginning in problems related to games of chance, it evolved haphazardly until the early nineteenth century when it emerged, almost fully grown, as the magnificent *Doctrine de Chance* of Laplace. It was then almost immediately lost in the ruthless march toward absolute rigor, which had begun at about that time. By the time our century came along, probability theory was not even considered to be part of mathematics. It was readmitted into the fold in the 1930s only after A. N. Kolmogorov provided a complete axiomatization and made the subject "respectable."

Probabilistic ideas entered the more traditional branches of mathematics slowly, and their impact, though steadily on the increase, has not been dramatic. It is mainly felt in the theory of parabolic and elliptic differential equations, and the intervention of probabilistic concepts comes through utilization of the idea of integration in function spaces.

Integration in function spaces was first introduced by Norbert Wiener in a series of papers written around 1923, in which he developed a wholly novel approach to the theory of Brownian motion of a free particle. The standard theory tells us how to calculate the probability that a Brownian particle moving along the x-axis and having started at $t = 0$ from $x = 0$ will be found at times $t_1, t_2, \ldots, t_n$ ($t_1 < t_2 < \ldots < t_n$) in prescribed intervals $(a_1, b_1), (a_2, b_2), \ldots, (a_n, b_n)$. If a large number N of identical Brownian particles are released at $t = 0$ from $x = 0$ then the probability in question will approximate well the proportion of particles which at the prescribed times will pass through the preassigned "gates." Wiener proposed to interpret the probabilities as that above not in terms of *statistics of particles* as I have just done, but in terms of *statistics of paths*. In other words, he considered all possible paths (starting from $x = 0$ at $t = 0$) which a Brownian particle can choose, and interpreted probabilities as *measures* in the space of these paths.

Now, how does one go about constructing a measure theory? For the sake of simplicity let us think of the plane. We start by choosing a family of "elementary sets" whose measures are postulated. In the case of the plane, the elementary sets could be rectangles. The measures assigned to them would be their areas as defined in elementary geometry. We now introduce the axiom of additivity to the effect that if *non-overlapping* sets $A_1, A_2, \ldots$ are "measurable" (i.e. somehow a non-negative measure can be assigned to each of them), then the union of the sets is also measurable and its measure is the sum of measures of the As:

$$m(A_1 \cup A_2 \cup A_3 \cup \ldots) = m(A_1) + m(A_2) + \ldots$$

We also need the axiom of complementarity to the effect that if A and B are measurable and A is included in B, then $B - A$ (the set of elements in B which are not in A) is measurable.

With these two axioms (rules) and with rectangles whose measures are given we can assign measures to extremely complicated sets. Once the measure is constructed, we can introduce the concept of integration.

We can now try to imitate this procedure in the space of paths. The rectangles are replaced by sets of paths which pass through preassigned intervals at prescribed times and the areas by probabilities assigned to these sets by the standard theory of Brownian motion. The axioms of additivity and complementarity are, of course, retained.

After overcoming a number of technical difficulties, one ends up with a theory of measure and integration. Remarkably enough this seemingly bizarre extension of the ordinary theory of areas and volumes has found many applications to problems of classical analysis and physics. For example, for a region R in Euclidean three-space, the measure (in the sense of Wiener) of the sets of paths emanating from a point p outside R, which at some time will hit R, is equal to the electrostatic potential at p corresponding to a charge distribution for which the potential on the boundary of R is normalized to unity.

Wiener's ideas were dramatically rediscovered by Richard Feynman in 1948 (actually in his Princeton doctoral dissertation in 1942, which was not published until 1948) in the context of non-relativistic quantum mechanics. Considering (for the sake of simplicity) a system of one degree of freedom (i.e. a particle) with the Lagrangian

$$L[x(\tau)] = \frac{m}{2} \left(\frac{dx}{d\tau} \right)^2 - V(x(\tau)) \ .$$

Feynman defines the quantum mechanical propagator

$$K(x \mid x; t)$$

as the integral

$$\int d(\text{path}) \exp\left\{ \frac{i}{h} S[x(\tau)] \right\}$$

over all paths which originate from x at $\tau = 0$ and end at x at $\tau = t$. S is the action

$$S[x(\tau)] = \int_0^t L[x(\tau)] d\tau$$

and h the Planck constant divided by 2π.

The definition of the integral over the paths is now much trickier owing to the imaginary i in the exponent, and much mathematical work remains to be done. Formally, the theory is in many respects analogous to a theory based on the Wiener integral.

Function space integrals (or path integrals, as they are often referred to) are at the center of much of contemporary work, both in mathematics and in physics. I see no slackening of activity in this area in the near or even not-so-near future.

I should like to mention one more example of the influence of probabilistic ideas, this time on one of the "purest" branches of mathematics, namely number theory. In 1939 Paul Erdös and I proved that the density of integers n whose number of prime divisors $\nu(n)$ satisfies the inequality

$$\log \log n + \alpha\sqrt{(\log \log n)} < \nu(n) < \log \log n + \beta\sqrt{(\log \log n)}$$

is equal to the error integral

$$\frac{1}{\sqrt{2\pi}} \int_\alpha^\beta \exp(-x^2/2) dx \ .$$

I regret that I am referring to my own work, but it happens

to be a good illustration of a point I am about to make. The difficult part of the proof (which was supplied by Erdös) involved purely number-theoretical arguments. But it was an argument based on probability theory which suggested the theorem.[3] If probability theory was at that time still "out of bounds" to mathematicians (and to many it was) there would have been no way of connecting prime divisors with the law of errors. On the other hand if in 1916 the Poisson distribution was not thought about as merely being empirically applicable to such things as the number of mule-kicks received by Prussian soldiers, then Hardy and Ramanujan could have discovered the theorem. As it is they came very close to it.

Brief, incomplete, and superficial as this account is, it must include at least a mention of combinatorial theory. Traditionally this part of mathematics, whose origins are buried in antiquity, has been concerned with the art and science of counting, enumerating, and classifying. It gradually annexed more and more ground, and so it has now become the study of all discrete and finite structures. Because of the enormity of its scope, embracing as it does the Ising model, the four-color problem, and finite projective geometries, as well as more familiar questions of counting (e.g. how many different necklaces of n beads of k colors are there?), it lacked unified methodology and clear-cut criteria whereby puzzles could be distinguished from questions of genuine mathematical or scientific interest. It thus stood until recently on the periphery of mathematics, attracting few practitioners and frightening many by the prohibitive difficulty of its problems.

In the past twenty years we have witnessed a profound change. A search for unifying principles as well as an attempt to move combinatorial theory into the mainstream of mathematics had begun and strides have already been made. In Chapter V of this book, G. C. Rota (one of the principal figures in the combinatorial renaissance) draws and paints an excellent picture

---

[3]For a popular discussion of probabilistic ideas in number theory see, e.g. [11].

of the subject and prophesizes an explosive future — a prophecy which to a large extent has been fulfilled.

Of the example discussed by Rota, I have chosen one because it helps to illustrate the mysterious ways in which mathematics sometimes moves.

The example concerns the so-called travelling salesman problem, which can be stated as follows: suppose that we have n cities and the cost of travelling from city i to city j is $c(i, j)$; how should one plan the route through all the cities which will minimize the total cost?

The problem is of a type encountered in economics under the general heading of allocation of resources. It seems totally lacking in mathematical interest, since there is only a finite number of possible routes and, correspondingly, a finite number of possible costs to compare. But now comes the crux of the matter, since what is at stake is how long would it take to reach the solution, and this is clearly the problem of estimating the algorithmic complexity of the task at hand.

It has been shown that this complexity is less than

$$c\, n^2\, 2^n \, ,$$

where c is a certain constant, but it is not known whether this is the right order of magnitude. If it were, the problem would have to be considered as unsolvable, since the amount of computing time needed to find the solution, even for moderate n, would become prohibitive. I conclude with a quotation from Rota's article:

> Attempts to solve the travelling salesman problem and related problems of discrete minimization have led to a revival and a great development of the theory of polyhedra in spaces of n dimensions, which lay practically untouched — except for isolated results — since Archimedes. Recent work has created a field of unsuspected beauty and power, which is far from being exhausted. Strangely, the combinatorial study of polyhedra turns out to have a close connection with topology, which is not yet understood. It is related also to the theory behind linear programming and similar methods widely used in business and economics.
>
> The idea we have sketched, of considering a problem $S(n)$ depending on an integer n as unsolvable if $S(n)$ grows too fast, occurs in much

the same way in an entirely different context, namely number theory. Current work on Hilbert's tenth problem (solving Diophantine equations in integers) relies on the same principle and uses similar techniques.

As we have stated earlier, Hilbert's tenth problem has in the meantime been solved.

I must end by apologizing for having left out large areas of mathematics which are vigorously pursued and are held in higher esteem by the community than the developments I have selected as being in some way indicative of the present state (good or bad) or possibly holding seeds of future concerns.

Certainly algebraic geometry, now much in vogue (two of the ten "peaks" selected by Halmos *et al.* are solutions of outstanding problems in this field), is much broader in scope and clearly of greater intrinsic value to mathematics than catastrophe theory. But algebraic geometry, like several other important branches which I have failed even to mention, is to a great extent *internal* to mathematics, and I have tried, to the best of my ability and judgement, to single out for emphasis themes and trends which may contribute to lessening of the isolation of mathematics.

To the many omissions I must add one more. In recent years there has been much progress in the field of non-linear differential equations. Results about singularities of solutions of field equations of general relativity are surely among the most striking advances of the past few years. The recently discovered method of solving a class of important non-linear equations by the inverse scattering method is a new chapter of classical analysis of great beauty and appeal. It also lifted solitons from hydrodynamic obscurity into objects of intensive study in both mathematics and physics (the fields ϕ of finite energy discussed briefly in § 7.3 are solitons). Last, but perhaps not least, the remarkable non-linear phenomena in so-called dissipative structures, though still in their mathematical infancy, will surely inspire much further work. All these topics deserve much more than being included in an apology for their omission.

Mathematics today is a vital, vibrant discipline composed of many parts which in mysterious ways influence and enrich each other (*im Dunkeln befruchten* to borrow a Faustian phrase from Hermann Weyl)[4]. It is beginning to emerge from a self-imposed isolation and listen with attention to the voices of nature. There is only one danger one must guard against, and that is that some zealots will, on either aesthetic or on counter-aesthetic grounds, try to redefine mathematics, so as to exclude this or that of its parts. Culture like nature has its ecological aspects, and the price for interfering with established equilibria may be catastrophically high.

1979

References

1. STONE, M. The revolution in mathematics. *American Mathematical Monthly* 68, 715-34 (1961).

2. EWING, J. H., GUSTAFSON, W. H. HALMOS, P. R., MOOLGAVKAR, S. H. WHEELER, W. H., and ZIEMER, W. P. American mathematics from 1940 to the day before yesterday. *American Mathematical Monthly* 83, 503-16 (1976).

3. McKEAN, H. P. and Singer, I. M. Curvature and the eigenvalues of the Laplacian. *Journal of Differential Geometry* 1, 43-69 (1967).

4. DEAKIN, M. A. B. Catastrophe theory and its applications. *Mathematical Scientist* 2, 73-94 (1977).

5. FOWLER, D. H. The Riemann-Hugoniot catastrophe theory and the van der Waals equation. In *Towards a theoretical biology*, Vol. 4, pp. 1-7. (Ed. V. H. Waddington.) Edinburgh University Press.

6. KAC, M. *Quelques problèmes mathématiques en physique statistique*. Les Presses de l'Université de Montréal (1974).

7. KAC, M. On applying mathematics: reflections and examples. *Quarterly Journal of Applied Mathematics* 30, 17-29 (1972).

8. CHAITIN, G. J. Randomness and mathematical proof. *Scientific American* 232 (5), 47-52 (1975).

9. MAY, R. M. Simple mathematical models with very complicated dynamics. *Nature* 261, 459-67 (1970).

10. MAY, R. M. Biological population nonoverlapping generations: stable points, stable cycles, and chaos. *Science* 186, 645-7 (1974).

11. LINNIK, Y. V. Les nombres entries se pretent-ils aux jeux hasard? *Atoms* 245, 441-6 (1967).

12. ROTA, G. C. Combinatorial analysis. In the present volume.

---

[1]See his *Vorwort* to his classic *Gruppentheorie und Quantummechanik.*

CHAPTER NINE

The Future of Computer Science

Till Now: Technology as the Driving Force

LET ME BEGIN WITH THE REMARK, taken from the recent NRC Study "Roles of Industry and University in Computer Research and Development 1," that two basic ideas, namely Babbage's stored program concept and its triumphant transistor/microcircuit implementation, continue after thirty years to drive the computer field as a whole. These ideas have proved so rich in consequence as to ensure the practical success of our field, independent of any other accomplishment. Since, as Turing emphasized, the Babbage/Von Neumann processor is computationally universal, any computational paradigm is accessible to it, so that it can be improved in speed and size only, never in fundamental capability. Nevertheless, the successes of microcircuit technology have enabled computer speed and size to grow in an astounding way, confronting workers in the field with a continually widening range of opportunities. For example, discs have led to databases, communication technology has led to ARPANET, and microchips to interest in distributed computing.

Based upon this technological/economic progress, the applications of computer technology have broadened rapidly, much more rapidly than the technology itself has deepened. Though constrained by software approaches that are barely acceptable, an extraordinary dissemination of computer technology has begun and will surely continue through the next decade.

The computer has become a household item, and besides its business applications, now stands at the center of a large entertainment industry. This guarantees the use of mass production techniques, implying yet further cost reductions. We can expect that a lush computational environment, represented by families of new personal computers whose performance will rapidly rise to several MIPS, will become commonplace for the scientist and business executive over the next few years, and with some additional delay will become common in homes. Continued improvements in magnetic recording densities, plus use of the extraordinary storage capacity of videodisk, will make large amounts of fixed and variable data available in the personal computing environment which we are beginning to see. Where advantageous, the computational power of workhorse microprocessor chips will be further buttressed by the use of specialized chips performing particular functions at extreme speeds. Chips for specialized graphic display and audio output functions are already available, and other such chips, performing signal analysis, data communication, data retrieval, and many other functions will be forthcoming. Finally, cheap processor chips make new generations of enormously powerful, highly parallel machines economically feasible.

The next few years will begin to reveal the fruits of these new opportunities. I believe, for example, that we will see a flowering of computer applications to education which will reflect our new ability to integrate microcomputers, computer graphics and text display, audio and video output, large videodisk databases, including stills and moving images, and touch-sensitive screens into truly wonderful, game-like learning environments. The video-game industry is surely alive to these possibilities, and I believe that its activity will affect both education and publishing more deeply than is generally suspected. All this will be part of the development of a world of personal and business information services more comfortable and attractive than anything we have yet seen.

Till Now: The Swamp of Complexity

Thus, hardware trends, reflecting the microcircuit engineer's astonishing progress, face us with the opportunity and requirement to construct systems of extraordinary functionality. However, even in the best of cases, this increased functionality will imply increased systems complexity. Moreover, since lack of firm understanding of how to proceed with software has never fully restrained the drive to increased functionality, huge, sprawling, internally disorganized and externally bewildering systems of overwhelming complexity have already been built, and have come to plague the computer industry. These dimly illuminated, leaky vessels force more of their crew to bail, and to grope about for half-hidden control switches, than are available to row forward. The situation is ameliorated in the short term, but at length exacerbated, by the economic "softness" of software: costs are incurred incrementally rather than in a concentrated way; no single detail is difficult in itself, but the accumulation of hundreds or thousands of discordant details makes the overall situation impossible. This is the "software crisis."

Language and software systems designers have attempted to find ways of slowing the growth of this sprawling disorganization, and have succeeded to at least some modest degree. First assemblers, then compilers, and then compilers for progressively higher levels of language have been developed to allow the pettiest and most onerous levels of detail to be handled automatically. This has helped overcome one of the major obstacles to stabilization and re-use of software, namely pervasive infection of codes by machine related details. This is part of the reason why the first generation of symbolic programming languages, specifically FORTRAN and COBOL, proved such a boon to the industry. Their success is shown by the large libraries of numerical, statistical, engineering, and other widely used applications programs that have accumulated in them. LISP has performed something of the same good service for workers in artificial intelligence. Substantially higher levels of language, which enable the programmer to use comfortable, human-like levels of abstraction somewhat more directly, also promise to make it easier to move major pieces of software from one application to another. It is fortunate that falling hardware costs are encouraging the use of these computationally expensive languages.

During the last few years, the technology of software transportability, exemplified by UNIX, various PASCAL compilers, and numerous microcomputer software systems has become well established. Emphasis on ideas of modularity and information hiding, about to receive a major test in the new DoD ADA language, should also contribute to the re-usability of software modules.

Better support tools have been provided to aid the programmer's work, and this improvement is continuing. Today's fast-turn-around environment, with an interactive editing, text-search facilities, and debugging displays is a substantial boon. However, much more remains to be done to integrate the programmer's editing/ compiling/debugging/tracing/version-control work environment into a truly helpful support system. This is a sprawling, complex problem, and I will not even begin the enumeration of its many facets. Let me note, however, that experimental development of more sophisticated program analysis tools, e.g. routines which pinpoint bugs by searching for textual incongruities in programs, is going forward and should lead to useful products.

Conceptual progress in the systems area has been linked to the discovery of families of objects, operations, and notations that work well together in particular application areas. Some of these object/operator families have proved to be of relatively general applicability, others to be powerful, but only in limited domains. Important examples are: the use of BNF grammars, variously annotated and "attributed," for the syntactic and semantic description of languages has been central to the development of "compiler-compilers." Tony Hoare's notion of "monitor" furnishes a significant conceptual tool for organizing the otherwise confusing internal structure of operating systems. The well-chosen family of string-matching operations of the

SNOBOL language, with the success/failure driven, automatically backtracking control environment in which they are embedded, allows many string-related applications to be programmed with remarkable elegance and efficiency.

Emergence of an Intellectual Precipitate Having Long-Term Significance

Though buffeted, and sometimes rendered obsolete by steadily changing technology, three decades of work by language and systems designers have given us products whose broad use testifies to a substantial degree of usability. However, since in this area it is so hard to distinguish real technical content from the arbitrary effects of corporate investment, salesmanship, and accidents of priority, it is gratifying that a solid theoretical inheritance has begun to emerge from this raw systems material. At the very least, this gives us an area in which the individualistic assertions of taste and the irresolvable debates characterizing the systems area yield to clear criteria of progress and theoretically based understanding of limits and tradeoffs. Whereas systems will change with technology, this work, like the mathematics it resembles, will be as important a century from now as it is today.

Central to this, the intellectual heart of computer science, stands the activity of algorithm design. By now, thousands of algorithmic clevernesses have been found and published. Of all the work that could be adduced, let me draw attention to four strands only. Many very useful and ingenious data structures, supporting important combinations of operations with remarkable efficiency, have been found. Hash tables, B-trees and the many variants of them that have been analyzed in the recent literature, and the specialized representations used to handle equivalence relationships efficiently are some of the prettiest discoveries in this subarea. Ingenious ways of arranging and ordering work so as to minimize it have been uncovered and exploited. The well-known divide-and-conquer paradigm, and also Bob Tarjan's many depth-first-spanning-tree based graph algorithms, exemplify this remark. Clever combinatorial reformulations have been found to permit efficient calculation of graph-related quantities which at first sight would appear to be very expensive computationally. The fast algorithms available for calculation of maximal matches in constrained-choice problems, maximum network flows, and also some of Tarjan's algorithms for solving path problems in graphs all illustrate this third point. Finally, deep-lying algebraic identities have been used to speed up important numerical and algebraic calculations, notably including the discrete Fourier transform and various matrix operations.

The techniques of formal algorithm performance analysis, and the fundamental notion of asymptotic performances of an algorithm, both put squarely on the intellectual map by Donald Knuth, give us a firm basis for understanding the significance of the algorithms listed in the preceding paragraph and the thousands of others that have appeared in the computer science literature. An algorithm that improves the best known asymptotic performance for a particular problem is as indisputably a discovery as a new mathematical theorem. This formal criterion of success, by now firmly established, has allowed the work of algorithm designers to go forward in a manner as systematic and free of dispute as the work of mathematicians, to which it is similar.

Studies of the ultimate limits of algorithm performance have deepened the work of the algorithm designer, by putting this work into a mathematical context drawing upon and deepening profound ideas taken from mathematical logic and abstract algebra. Landmarks here are the Cook-Karp notion of NP-completeness, which has proved to draw a very useful line between the feasible and infeasible in many algorithmic areas close to practice; the work on exponentially difficult problems deriving from Meyer, Stockmeyer, and Rabin; and various more recent investigations, by "pebbling" and other ingenious combinatorial methods, of time-space tradeoffs in computation.

The mathematical affinities of most of this work are with mathematical logic, combinatorics, and abstract algebra, which till now have been the branches of mathematics of most concern to the computer scientist. However, developments now afoot seem likely to renew the ties to classical applied mathematics which were so important in the earliest days of computing.

Trends

Some of the dominant research currents of the next decade will merely deepen what is already important, others will bring matters now attracting relatively limited attention to center stage. The VLSI revolution will surely continue, dropping the price of processing and storage. As already stated, this ensures that most of the existing army of programmers will be occupied in keeping up with the developmental opportunities that this technology will continue to provide. To avoid complete paralysis through endless rework of existing software for new hardware, software approaches which stress transportability and carry-forward of major systems, plus re-usability of significant software submodules, must and will become more firmly established. Growing understanding of the most commonly required software subfunctions should allow many of them to be embodied in broadly usable software, firmware, or hardware packages. Besides numerical, statistical, and engineering packages of this sort, operating system, database, communication, graphics, and screen management packages should all emerge.

The steadily broadening applications of computers will draw algorithm development into new areas. Computational geometry, i.e. the design of efficient ways for performing many sorts of geometric computations, has flourished greatly since its first buddings a few years ago and gives a good indication of what to expect. This area of algorithmics contributes to VLSI design, CAD/CAM development, and robotics, all of which have nourished it.

Robotics can be expected to play a particularly large role in bringing new directions of investigation to the attention of algorithm designers. Though it may be expected to draw upon many of the other branches of computer science, robotics will have a different flavor than any of them, because in robotics, computer science must go beyond the numeric combinatorial, and symbolic manipulations that have been its main concern till now: it must confront the geometric, dynamic, and physical realities of three-dimensional space. This confrontation can be expected to generate a great deal of new science, and will greatly deepen the connections of computer science to classical applied mathematics and physics.

If, as I expect, robotics becomes as central to the next few decades of computer science research as language, compiler, and system-related investigations have been during the past two decades, significant educational and curricular problems will confront many computer science departments. Computer science students' knowledge of the basics of applied mathematics and physics is small at present, and I suspect that it is diminishing as the systems curriculum expands and older connections with mathematics are lost. For roboticists, this trend will have to be reversed: solid training in calculus and differential equations, linear algebra, numerical analysis, basic physics, mechanics, and perhaps even control theory will be necessary. Students going on to graduate work and research careers will also have to concern themselves with computational algebra, computational geometry, theories of elasticity and friction, sensor physics, materials science, and manufacturing technology.

Beyond the educational problem of how to squeeze all of this material into an already crowded curriculum, administrative problems of concern to sponsoring agencies will also arise. Presently, computer acquisition is regarded as a capital investment: once acquired, processors, storage units, terminals, printers are expected to remain in productive service over an extended period and to be used by many researchers for many purposes. Though some robotic equipment will have this same character, other equally important (and expensive) equipment items will, like the equipment of the experimental physicist, be useful only for specific, limited sequences of experiments. Robotics researchers will therefore have to learn how to define experiments likely to produce results worth high concentrated costs, and funding agencies will have to learn how to pay bills of a kind new to computer science.

One last area of practical and theoretical development deserves mention. Large-scale parallel computers seem certain to appear during the next few years as superperformance supplements to the ubiquitous microchip. Many university groups are currently preparing designs of this sort, and industry, not excluding Japanese industry, is beginning to become interested. This appears certain to make parallel algorithm design and the computational complexity theory of such algorithms active research areas during the next few years.

Eldorado

Though presently revolving around technology and such prosaic accomplishments as putting a word-processor in every office and home, the computer development leads forward to something immensely significant: the development of artificial intelligences which can at first match, and perhaps soon thereafter greatly surpass, human intelligence itself. The belief that this is possible motivates many computer scientists, whether or not they concern themselves directly with this subfield of computer science. It would be hard to overestimate the consequences of such an outcome.

Since this is the deepest perspective of our field, I shall conclude this talk by addressing it briefly. Concerning artificial intelligence, I think it well to comment with high hope concerning its long-term future, skeptically concerning its accomplishments to date, and in repudiation of the relentless overselling which has surrounded it, which I believe discourages those sober scientific discernments which are essential to the real progress of the subject.

Let me begin by subscribing to the materialist, reductionist philosophy characteristic of most workers in this field. I join them in believing that intelligent machines are indeed possible, since like them I view the human brain as a digital information processing engine, a "meat machine" in Marvin Minsky's aptly coarse phrase, wonderfully complex and subtle though the brain may be. This said, however, we must still come to grips with the main problem and dilemma of the field, which can, I believe, be formulated roughly as follows. To communicate with a computer, one must at present *program,* i.e. supply highly structured masses of commands and data. Communications with other persons is much easier, since they can digest relatively unstructured information, and can supply the still elusive steps of error correction and integration needed to use such information successfully. Thus only to the extent that a computer can absorb fragmented material and organize it into useful patterns can we properly speak of it as having "intelligence."

Researchers in artificial intelligence have therefore sought general principles which, supplied as part of a computer's initial endowment, would permit a substantial degree of self-organization thereafter. Various formal devices allowing useful structures to be distilled from masses of disjointed information have been considered as candidates. These include graph search, deduction from predicate algorithms, and in today's misnamed "expert" systems, the use of control structures driven by the progressive evaluation of a multiparameter functions whose parameters are initially unknown.

The hoped-for, and in part real, advantages of each of these formal approaches have triggered successive waves of enthusiasm and each approach has lent particular terminological color to at least one period of work in artificial intelligence. Any collection of transformations acting on a common family of states defines a graph, and discovery of a sequence of transformations which carries from a given to a desired state can therefore be viewed as a problem in graph search. Accordingly, transformation of graph into path can be regarded as a general formal mechanism which permits one to derive something structured, to wit a path, from something unstructured, namely its underlying graph. Hope that this principle would be universally effective characterized the first period of work in artificial intelligence, embodied in systems such as the "General Problem Solver," and in languages such as Planner.

The relationship of a proof in predicate calculus to the order-free collection of axioms on which this proof is based gives a second principle allowing ordered structures (i.e. proofs) to arise from unordered items of information (i.e. axioms). Because of its deep connections to mathematical logic and its great generality, this formal approach seemed particularly promising, and in its turn encouraged over a decade of work on resolution proof methods and their application. Today a new approach stands at the center of attention. Though considerably less deep and general than either the graph search or the predicate techniques which preceded it, this ask-and-evaluate sequencing paradigm, much used in today's "expert" consultant systems, has found enthusiastic advocates.

Sustained and patient work aimed at full elucidation of the real, albeit very partial, insights inherent in each of these approaches is certainly appropriate. All, however, face a common problem: their efficiency decreases, sometimes at a catastrophic rate, as they are required to deal with more information and as the situations which they try to handle are made more realistic. It is well-known, for example, that computerized predicate logic systems which can easily derive a theorem from twelve carefully chosen axioms will often begin to flounder badly if three more axioms are added to these twelve, and will generally fail completely if its initial endowment of axioms is raised to several dozen. Similarly, graph search methods that work beautifully for monkey-and-bananas tests involving a few dozen or hundred nodes fail completely when applied to the immense graphs needed to represent serious combinatorial or symbolic problems. For this reason I believe it very premature to speak of "knowledge-based" systems, since the performance of all available systems degenerates if the mass of information presented to them is enlarged without careful prestructuring. That this is so should come as no surprise, since, not knowing how to specialize appropriately, we deal here with mechanisms general enough to be subject to theoretically derived assertions concerning exponentially rising minimal algorithmic cost. This observation reminds us that work in artifical intelligence must learn to reckon more adequately with the presence of this absolute barrier, which stubborn over-optimism has till now preferred to ignore.

Then along what lines is progress toward the goal of artificial intelligence to be expected? Through incremental, catch-as-catch-can improvement of our capacity to mimic individual components of intelligent function, such as visual pattern recognition, speech decoding, motion control, etc.? By integrating algorithmic methods into systems which use the traditional A.I. techniques only as a last resort? Through eventual discovery of the A.I. philosopher's stone, namely some principle of self organization which is at once effective and efficient in all significant cases? By applying vastly increased computing power? We do not know, but the attempt to find out will continue to constitute the most profound, though not necessarily the most immediate, research challenge to computer science over the coming decades.

1978

CHAPTER TEN

Economics, Mathematical and Empirical

THOUGH ITS INFLUENCE IS A HIDDEN ONE, mathematics has shaped our world in fundamental ways. What is the practical value of mathematics? What would be lost if we had access only to common sense reasoning? A three-part answer can be attempted:

1. Because of mathematics' precise, formal character, mathematical arguments *remain sound even if they are long and complex.* In contrast, common sense arguments can generally be trusted only if they remain short; even moderately long nonmathematical arguments rapidly become far-fetched and dubious.

2. The precisely defined formalisms of mathematics discipline mathematical reasoning, and thus *stake out an area within which patterns of reasoning have a reproducible, objective character.* Since mathematics defines the notion of formal proof precisely enough for the correctness of a fully detailed proof to be verified mechanically (for example, by a computing mechanism), doubt as to whether a given theorem has or has not been proved (from a given set of assumptions) can never persist for long. This enables mathematicians to work as a unified international community, whose reasoning transcends national boundaries and survives the rise and fall of religions and of empires.

3. Though founded on a very small set of fundamental formal principles, mathematics *allows an immense variety of intellectual structures to be elaborated.* Mathematics can be regarded as a grammar defining a language in which a very wide range of themes can

be expressed. Thus mathematics provides a disciplined mechanism for devising frameworks into which the facts provided by empirical science will fit and within which originally disparate facts will buttress one another and rise above the primitively empirical. Rather than hampering the reach of imagination, the formal discipline inherent in mathematics often allows one to exploit connections which the untutored thinker would have feared to use. Common sense is imprisoned in a relatively narrow range of intuitions. Mathematics, because it can guarantee the validity of lengthier and more complex reasonings than common sense admits, is able to break out of this confinement, transcending and correcting intuition.

By fitting originally disparate theories to the common mold defined by the grammar which is mathematics, one gives these theories a uniform base, which among other things makes it possible for theories which initially stand in irreducibly hostile confrontation to be related and compared, perhaps by means of some comprehensive mathematical construction which includes all these theories as special cases.

On the other hand, the precise character of the axiomatic formalisms of mathematics is attained at the cost of a radical divorce between these formalisms and the real world. Therefore, in regard to real-world applications, the mathematical mode of reasoning will only retain its advantage if it is based upon some very compelling model of an aspect of the real world. Only empirical work whose direct confrontation with reality lies outside mathematics itself can validate such a model, though part of the inspiration for such models may come from within mathematics.

It is sobering to note that we possess only two techniques for ensuring the objective validity of our thought. The approach of empirical science — to anchor oneself directly to verifiable statements concerning real-world observations — is one of these techniques; and mathematics embodies the other. Science disciplines thought through contact with reality, mathematics by adhering to a set of formal rules which, though fixed, allow an endless variety of intellectual structures to be elaborated.

The external disciplines which science and mathematics impose upon their practitioners play particularly essential roles in those cases in which the truth is surprising or in which some captivating established doctrine needs to be broken down. Common sense, acting by an undisciplined principle of plausibility and intuitive appeal, cannot cope either with situations where truth is even partly counterintuitive, or with situations in which ingrained familiarity has lent overwhelming plausibility to some established error. Thus, for example, common sense can cope adequately with situations where, by shoving an object in a given direction, one causes it to move in that direction; but in those more complicated situations in which objects move in directions contrary to the dictates of intuition, the more delicate techniques of formal mathematical reasoning become indispensable.

However, application of the cautious procedures of science and of mathematics proceeds slowly, whereas the daily affairs of hurrying mankind demand constant resolution. Thus, where adequate knowledge and objective theory do not exist, fragmentary heaps of fact and primitive common sense reasoning will be drawn upon to fill the gap. For this reason, both personal decisions and social commitments of vast consequence often rest on doubtful and inadequate understanding. Moreover, just as a person committed to a given course may resist and resent efforts to point out irrational aspects of his behavior, so societies can resent and even repress criticism of intellectual structures, correct or not, which are regarded as essential to a social consensus and around which significant interests may have collected. In such situations, mathematics has often been used as a tool for defusing controversy and for allowing ideas running counter to some dominant doctrine to be developed in a neutral form.

In the present article we will try to illustrate the way in which mathematics can contribute to our understanding of social issues. In order to do this, and specifically in order to make the point that mathematical reasoning need by no means remain remote and abstract, we will venture to reason about hard-fought areas of current controversy. This makes it all the more necessary for the reader to remember that our reasoning, like all applications of mathematics to the real world, is not at all incontrovertible, but merely represents a set of conclusions to which one can be led by assuming one particular model of reality. Any such model is but one of many possible models of differing adequacy which can lead to different, and sometimes diametrically conflicting deductions. Our reasoning can therefore not tell us that our conclusions concerning reality are correct, but only that they are logically possible.

Some Classical Economic Issues

Economics, the body of theory concerned with the mechanisms by which society pushes its members to productive activity and allocates the fruits of that activity, illustrates the points made in the preceding paragraphs. Since economic institutions must link men into extensive patterns of joint action, and since these patterns must be forged in an area in which conflicting individual and group claims are never very far from the surface, no society is able to do without some unifying and generally accepted economic ideology. The economic ideology of our own American society generally emphasizes the importance of unhampered individual and corporate enterprise in production, and of individual choice in consumption. However, there also exist centralized legal mechanisms which forbid certain types of production (e.g., of unpasteurized milk except under special circumstances, use of child labor in factories) and consumption (e.g., narcotic drugs except under special circumstances). In fact, the range of coercive governmental interventions which now modify the workings of the decentralized economic machine is quite broad. For example, the government regulates allocation of the economic product among those who have cooperated in its production (e.g., by establishing minimum wage standards) and among major social subgroups (through taxing patterns and welfare programs); directly influences vital economic parameters (e.g., availability of credit, through bank regulation and the activity of the central banking system); prevents certain types of economic cooperation and encourages others (through anti-trust and labor relations laws); and prohibits certain types of misrepresentation and irresponsibility (e.g., through laws which punish commercial fraud, demand certain forms of record-keeping and information disclosure and compel individuals to contribute to a central pension fund).

To make any one of these numerous economic interventions attractive to the public and acceptable to the legislator, there is required some fragment of theory, which, whether it be primitive or sophisticated, correct or mistaken, must at least be persuasive. To commit society as a whole to an overall economic approach, a more systematic body of theory recommending this approach on moral and pragmatic grounds is required. For this reason, the stock of available economic theories has exerted a fundamental influence on American and world society. The classical case for a resolutely *laissez-faire* approach is that made in Adam Smith's *Wealth of Nations:*

> ... As every individual, therefore, endeavors as much as he can both to employ his capital in support of domestic industry, and so to direct that industry that its produce may be of greatest value, every individual necessarily labors to render the annual revenue of the society as great as he can. He generally, indeed, neither intends to promote the public interest, nor knows how much he is promoting it. By preferring the support of domestic to that of foreign industry, he intends only his own security; and by directing that industry in such a manner as its produce may be of the greatest value, he intends only his own gain, and he is in this, as in many other cases, led by an invisible hand to promote an end which was no part of his intention.

Since their publication in 1776, these words of Adam Smith have been immensely influential, and have been repeated and refined by subsequent generations of market-oriented economic thinkers. Of course, arguments supporting entirely different conclusions would not be hard to cite.

How can we begin to find safe ground from which to judge such arguments? To begin with, it is well to note that these arguments are only justified to the extent that the pragmatic judgment implicit in them is in fact correct. If Adam Smith is correct in asserting that his "invisible hand" acts to harmonize interactions between the participants in a free economy, then his recommendations are wise; but if these interactions cause the participants in such an economy to act at crosspurposes and to frustrate each other's intentions, then he is mistaken. This implies that the free-market arguments of Smith and his disciples, like the arguments for particular measures of centralized control advanced by various of his ideological opponents, have substance only insofar as more penetrating analysis is able to establish the extent to which an economy organized on the basis of particular freedoms or controls will actually deliver to its members the goods that they want.

Thus what one needs to analyze these arguments is a neutral theoretical framework within which the interaction of individuals in an economic system can be modeled, and from which some account of the results of their interaction can be developed. In principle, the input to such a model will be a mass of rules and statistical data describing the personal, technological, and legal facts which constrain the participants, and the approaches which they typically employ in their effort to adapt to shifting economic circumstances. The model must be broad enough to allow incorporation of any element of motivation or economic response deemed significant, and should also be general and "neutral," i.e., free of preconceptions of the result that will emerge from the multiparticipant interactions being modeled.

Mathematical Models of Social and Economic Interaction

Mathematical models within which a national economy could be described by a closed self-determining system of equations began to appear in the 1870s. The limited class of models initially proposed (notably by Leon Walras in Switzerland and Vilfredo Pareto in Italy) was substantially and dramatically broadened in the late 1920s and early 1930s in a series of papers by the brilliant and universal twentieth century mathematician John von Neumann (1903-1957), whose work on these matters was rounded out and made generally available in the treatise *Theory of Games and Economic Behavior* published in 1944 by von

Neumann and the late Oscar Morgenstern. The essential accomplishment of von Neumann and Morgenstern was to imbed the limited supply/demand models which had previously been considered into a much larger class of models within which substantially more varied patterns of interpersonal relationship could be brought under mathematical scrutiny.

In a model of the von Neumann-Morgenstern type, one assumes a finite number of participants, whose interactions are constrained (but not uniquely determined) by a set of precisely defined, prespecified rules. In the economic application of models of this sort, the participants represent producers and consumers, and the rules describe the legal and technical facts which constrain their economic behavior. In a sociological or political-science application of a von Neumann-Morgenstern model, participants might either be individuals or nations, and the rules of the model would then describe the institutional, traditional, or military factors defining the options open to them and the consequences which will follow from the exercise of a given pattern of options. It is useful to note that in addition to these very real sources of situations to be modeled, a stimulating variety of miniature situations readily describable by von Neumann models is furnished by the large collection of familiar board, card, and nonathletic childen's games (such as Choosing or Stone-Paper-Scissors) with which everybody is familiar. It is because these games are so suggestive of directions of analysis that von Neumann and Morgenstern used the word "Games" in the title of their book and fell into the habit of referring to their models as games and to the participants whose interaction was being modeled as players.

The formal structure of a von Neumann-Morgenstern model makes available, to each of the participants or players which it assumes, some finite number of "moves" (which might also be called "policies" or "strategies") among which the participants are allowed to choose freely. It may be that one single set of moves is available to all the participants assumed by a model; on the other hand, the model may allow different sets of moves to different participants, in this way modeling real-world situations in which participants interact in an inherently unsymmetric way. After each player has made his move, a set of rewards (and penalties, which are regarded simply as negative rewards) are distributed to the players. In a game situation, we can think of these rewards or penalties as being distributed by an umpire who administers the rules of the game. In a more realistic situation we can think of these rewards and penalties as describing the actual outcome for each participant of the policies followed by him and by all other participants. These rewards or penalties may, for example, be business gains or losses, given the pattern in which other participants choose to invest, buy, and sell; or they may be military victories or defeats, given the pattern in which other participants choose to fight, remain neutral, or flee. In the terminology introduced by von Neumann and Morgenstern, the reward/penalty assessed against each participant in a game is that participant's "payoff."

Suppose that players $1, 2, \ldots, n$ respectively choose moves $m_1, m_2, \ldots, m_n$ out of the collection of moves available to them. It is the collective pattern of all these moves that determines the individual payoff to each player; thus the payoff to the jth player is some mathematical function of the whole list of variables $m_1, \ldots, m_n$. Therefore if we use the standard notation for mathematical functions we can write the payoff to the jth player as $p_j(m_1, m_2, \ldots, m_n)$. The payoff functions p_j summarize all those aspects of a game or other interactive situation which a particular model is able to capture, so that we may say that each particular von Neumann model is fully defined by such a set of functions.

The participants in a von Neumann-Morgenstern model are driven to interact simply because the payoff to the jth player is determined not only by his own move m_j, but also by the moves of all other participants. In selecting his move as a game is repeatedly played, each participant must therefore take account of the observed behavior of the other participants, and, if his own activity is a large enough factor in the total situation to affect these other participants significantly, must analyze the effect of his move on these other participants, and take account of the countermoves which they might make to repair their position if it is significantly changed by his choice of move. The following boxes illustrate the rich variety of models encompassed by this theory.

Bluff

Bluff is a two-player game with matching gains and losses. Player 1 writes a number from 1 to 10 on a slip of paper. Without showing this slip to Player 2, he tells Player 2 what he has written, but he is allowed to lie, announcing either the same number that he wrote down, or a different number. Player 2 must then guess whether Player 1 is telling the truth or not; he does so by saying true or lie. If caught in a lie, Player 1 must pay Player 2 a dollar; if falsely accused of lying, Player 1 collects half a dollar from Player 2. If Player 1 does not lie and Player 2 correctly guesses that he has not lied, then Player 1 must pay 10 cents to Player 2. If Player 1 lies without Player 2 guessing that he has done so, then he collects fifty cents from Player 2. These payoffs are summarized by the following payoff table for Player 1:

	m_2	
m_1	**true**	**lie**
same	-10	$+50$
different	$+50$	-100

Player 2's payoffs are the exact opposite: what Player 1 wins, Player 2 loses, and vice versa. This game is, therefore, an example of a two-person zero-sum game.

Cooperation

This is a two-player game in which interplayer cooperation is possible. Both players write either a 1 or a 2 on a slip of paper. If one of the players has written 2 on his slip while the other has written 1, then the umpire gives two dollars to the player who has written 2 and collects one dollar from the other player. If both players have written 1, then the umpire collects one dollar from each of the players. If both players have written 2, then the umpire collects ninety cents from both players. (The reader may wish to work out the best way for the players to cooperate.)

Litter

This game is a many-player model of the dropping of litter in streets. There are a large number of participants, perhaps 1000. Each participant is given a slip of paper which he is allowed to either dispose of, or drop. If a player drops his piece of paper, he must then roll a pair of dice; if the dice come up snake eyes (both ones) he must pay one dollar to the umpire. (This rule is intended to model the effect of an antilittering law which prescribes a fine of $1 for littering; we are in effect assuming that the probability of being caught by a policeman in the act of littering is 1/36, since 1/36 is the probability that two six-sided dice both come up 1.) If a player disposes of his piece of paper, he must pay 5¢ to the umpire. (This rule, which suggests a kind of garbage disposal service charge, can also be thought of as assigning a nominal cash value to the labor needed to dispose of an item of litter in a publicly provided receptacle.) After all n participants have chosen their move, the number of slips which have been dropped rather than disposed of is counted, and each participant is required (irrespective of whether he personally has disposed of or dropped his own slip of paper), to pay 1/10¢ to the umpire for each dropped slip. (This last rule can be thought of as assigning a nominal negative cash value to the degradation of the aesthetic environment caused by the accumulation of litter.)

Stagnation

This game is a many-participant majority game which models an everyday situation of stagnation. The model can have any number of participants; but let us suppose, to fix our thoughts, that there are ten. The participants desire to go in a group to receive some reward (e.g., to dine together). Each player can elect one of two moves, which we shall call *go* and *stay*. If more than half the players choose the move *go*, then those who choose the move *go* receive a payoff of one dollar from the umpire; in this case, those who choose the move *stay* receive nothing. If less than half the players choose the move *go*, then those who choose the move *go* must pay a fine of 10 cents, while those who choose the move *stay* pay nothing. (This last rule may be thought of as assigning a nominal cash value to the embarrassment which participants breaking away from a social majority may feel, and to the trouble which they are put to in rejoining the majority.)

Strategies and Coalitions

After introducing the broad class of models described above, von Neumann and Morgenstern entered into an analysis of the behavior of participants in these models, an analysis which they were able to carry surprisingly far. The simplest cases to analyze, and those in which the most conclusive analysis is possible, are those known as "two person zero-sum games." In these games one participant's gains are exactly the losses of the other participant, so that the two participants stand in a pure relationship of opposition, no sensible grounds for cooperation existing. (Mathematically, these are the games in which the two payoff functions p_1, p_2 satisfy $p_1 + p_2 = 0$, or, equivalently, $p_1 = -p_2$. This allows the whole game to be described by a single "payoff matrix" $p(m_1, m_2) = p_1(m_1, m_2)$.) For games of this type, von Neumann and Morgenstern were able to identify a clear and compelling notion of "best defensive strategy" which they showed would always exist, provided that the notion of move was broadened to allow for an appropriate element of randomization.

A simple yet instructive example is furnished by the familiar Stone-Paper-Scissors choosing game of children, in which each of the two participants independently chooses one of three possible moves, designated as "stone," "paper," "scissors," respectively. If both make the same choice, the play ends in a tie with no winner or loser. Otherwise the winner is determined by the rule "paper covers stone, stone breaks scissors, scissors cut paper." If we take the value of a win to be $+1$ and the cost of a loss to be -1 (while ties are of course represented by 0), then the payoff matrix for this game is as given in Figure 1.

	Player 1		
Player 2	Stone	Paper	Scissors
Stone	0	+1	−1
Paper	−1	0	+1
Scissors	+1	−1	0

(Payoff Matrix)

Move m	Best Response $R(m)$
Stone	Paper
Paper	Scissors
Scissors	Stone

(Best Response Function)

Figure 1. Stone-Paper-Scissors best response function.

A *perfect defensive move* in any game is a move which a player will find it advantageous to repeat, even assuming that this move comes to be expected by his opponent, who then replies with the most effective riposte available to him. So if either player makes a best defensive move and the other makes his best reply, then both players will simultaneously be making best defensive moves and best replies to the other's move. Suppose that we designate Player 2's best response to each possible move m_1 of Player 1 by $R_2(m_1)$, and symmetrically use $R_1(m_2)$ to designate Player 1's best response to Player 2's move m_2. Then if m_1^* and

m_2^* denote the corresponding perfect defensive moves, it follows that

$$m_1^* = R_1(m_2^*) \quad \text{and} \quad m_2^* = R_2(m_1^*) .$$

Hence to determine Player 1's best defensive move we have only to solve the equation $m_1^* = R_1(R_2(m_1^*))$.

In Stone-Paper-Scissors the payoff matrix is symmetrical, so the best response is the same for both players. In other words, $R_1 = R_2$, so we shall simply call it R; the best responses in this game are tabulated in Figure 1. Clearly, $R(R(\text{Stone}))$, the best response to the most effective riposte to Stone, is Scissors. Similarly, $R(R(\text{Paper}))$ is Stone, and $R(R(\text{Scissors}))$ is Paper. Thus in this game we find to our disappointment that there is no best defensive strategy (since there is no solution to the equation $R(R(m_1^*)) = m_1^*$).

However, as von Neumann was first to realize, this problem will always disappear if we allow players to *randomize* their moves, i.e., to assign fixed probabilities, totalling to 1, to the elementary (or pure) moves open to them according to the initially stated rules of a game, and then to make specific moves randomly but with these fixed probabilities. The necessary role of randomization is particularly obvious in games like Choosing, Penny-Matching, or Stone-Paper-Scissors, since it is perfectly clear that a player who always plays the same move is playing badly (no matter what single move he fixes on); to play well, a player must clearly randomize his moves in some appropriate fashion. Because of the perfect symmetry of the three options in Stone-Paper-Scissors, it is easily seen that the best randomization strategy in this game is to employ each move with equal probability.

The symmetric Stone-Paper-Scissors game offers no special advantage to either player. Other two-person zero-sum games are not so precisely balanced. The game Bluff described in the box on p. 125, for example, yields an average gain of $50/7 = 7.1$ cents per game to the first player, providing both participants play their best (randomized) strategy. (Details of these best defensive strategies are outlined in the box on p. 130.)

in a zero-sum game, one person's gain is another person's loss, so the second player loses, on average, an amount equal to the first person's gain; this amount is known as the "value" of the game. In general, the value of a two-person game is the amount won by one of the players if both players use their best defensive strategies. Thus a two-person game's value measures the extent to which the game inherently favors one or the other side when two players (or perhaps two coalitions of players) face each other in a situation of pure opposition.

Having established these fundamental facts concerning the simplest two-person games, von Neumann and Morgenstern proceeded to an analysis of n-person games. Their main idea was to consider all the coalitions that could form among subsets of the participants in such games, and to analyze the "strength" which each such coalition would be able to exert, and the extent to which each coalition could stabilize itself by distributing rewards among its participants, thus countering appeals for cooperation and promises of reward that might be made by other competing coalitions.

Best Bluffing Strategy

We analyze the game of Bluff (p. 125) by calculating best response functions. If Player 1 plays the move "same" with probability s and "different" with probability d, then Player 2 will gain an average of $10s - 50d$ if he plays "true," and $-50s + 100d$ if he plays "lie;" the relative advantage of "lie" is therefore $150d - 60s$. Hence player 2's best response to strategy is "lie" if $150d \geq 60s$, and "true" if $150d \leq 60s$. It follows that the best defensive strategy for Player 1 is to take $150d = 60s$, i.e., $d = 2/7$, $s = 5/7$; and similarly the best defensive strategy for Player 2 is to randomize by selecting "true" 5/7 of the time and "lie" the remaining 2/7. If both players of this game play their best strategies, the average gain of Player 1 is $-10(5/7) + 50(2/7) = 50/7 = 7.1$ cents per game (and of course the average loss of Player 2 is the same). We therefore say that the value of this game to Player 1 is $50/7$ (and to Player 2 is $-50/7$).

If we consider any subset C of the players of a game as a potential coalition, then it follows from the preceding analysis of two-person games that by agreeing among themselves on what their moves will be the members of C can secure for themselves a total amount $v(C)$ equal to the value of the two-person (or two-coalition) game in which all the members of C play together as a single corporate individual against all the remaining players taken as a single corporate opponent. This amount $v(C)$ is then available collectively to the members of C for distribution in some pattern chosen to stabilize the coalition, and measures the strength of the coalition C.

The coalitional analysis of von Neumann and Morgenstern begins with a study of the prototypical case where the number of players is three. In this case they were able to show that every "strategically interesting" game is equivalent to the game Choose-an-Ally in which each of three participants selects one of the two others as a partner. Then any two players who have designated each other collect a fixed sum from the third player and divide their gain. Clearly, in this game no strategy is possible other than the formation of two-person coalitions (who agree on their moves, though of course these agreements lie outside the strict rules of the game itself). With four participants, a three-parameter family of essentially different coalition strength patterns exists, so the analysis of coalitions is already intricate; the complexity of this analysis increases rapidly as the number of participants rises above four.

The coalitional analysis of von Neumann and Morgenstern uncovers a rich variety of phenomena. Nevertheless, we must assert that the direction in which this analysis proceeds is not the direction which is most appropriate for applications to economics. The reason is this: a coalitional theory of games rests on the assumption that the participants in the situation being modeled have the time and ability to work out agreements which lie entirely outside the defining rules of the game, and that these outside agreements are enforceable at least *de facto*, i.e., that once made they will be adhered to. Quite the opposite assumption, namely that the players are forbidden to or unable to communicate with each other except through the moves they make, is just as reasonable in many situations and is often much to be preferred. In particular, as von Neumann and Morgenstern point out, significant applications of their theory to economics depend on the treatment of *n*-person games for large *n*. But it is obvious that the maintenance and internal administration of coalitions of large numbers of members cannot be simple, and hence we may expect that, as the number of participants in a game increases, the tendency of the participants to be guided solely by the inherent constraints of the situation in which they find themselves and by observation of the moves made by other participants can become decisive.

Coalition-Free Equilibria

To understand what will happen in situations in which this is the case, what we need to do is to abandon coalition analysis as a false start, and to return to something very close to the original von Neumann notion of perfect defensive move, which has a direct and felicitous generalization to the *n*-participant situation. Specifically, we define a pattern $m_1^*, \ldots, m_n^*$ of moves in an *n*-player game to be an "equilibrium" if for each j, the move m_j^* of this jth player is the move which he finds it advantageous to make, given that the other players make the moves $m_1^*, \ldots, m_{j-1}^*, m_{j+1}^*, \ldots, m_n^*$. Such a (mutually confirming) pattern of moves will therefore give the jth player maximum payoff, assuming that he can count on the other players to make their equilibrium moves $m_1^*, \ldots, m_n^*$. Just as in the two-person game, the equilibrium pattern of moves is one in which every player is responding to the observed moves of all the other players in a manner which gives him maximum payoff.

This coalition-free viewpoint in game theory was first advanced seven years after the work of von Neumann and Morgenstern by the young American mathematician John Nash, and published by him under the title *Non-Cooperative Games*. The Nash approach opens the way to an effective treatment of *n*-person games for large *n*, and has the very appealing property of being based on an assumption whose validity should grow as *n* becomes large; by contrast, the von Neumann-Morgenstern coalitional approach founders for large n among rapidly growing combinatorial particularities.

From the economic point of view, the Nash equilibrium notion is particularly appealing since it can be regarded as a description of the equilibrium to which a game will move if the game's players are confined *de jure* or *de facto* by the normal rules of a "free market," namely by an inability to communicate outside the framework of the game, or by a legal prohibition against the formation of coalitions or cartels. From this angle we may say that whereas the equilibrium-point concept assumes that each player takes the context in which he finds himself as given and makes a personally optimal adjustment to this context, the von Neumann-Morgenstern coalitional analysis tends always to call the social constitution into question by asking whether any group within the economy is strong enough to force a drastic change in the established pattern. Thus, if one takes a view of economics rigorously consistent with the von Neumann--Morgenstern theory, economic analysis would necessarily include a theory of such things as Congressional negotiations on fiscal, tax, and tariff policy. However interesting such questions may be, their answers do not belong to economic theory in its most classical sense.

Having found all this to say in favor of coalition-free analysis, it is now appropriate to apply it. This will be easiest if we confine our attention to the rather large class of symmetric games, i.e., to games in which all players have essentially the same payoff function. Even though equilibria in which different players make different moves are perfectly possible even for such games, we shall for simplicity confine our attention to "symmetric equilibria" in which all players make the same move. These equilibria are determined by the equation $m^* = R(m^*)$, where $R(m)$ defines the best response of any player to a situation in which the move made by all other players is the same and is m.

Here, every participant finds the move m^* advantageous both because this move secures him a maximum payoff (given that all other players are making this same move, a fact which lies outside his control) and also because he finds the wisdom of his move confirmed by the fact that every other participant is observed to be doing the same thing. In such a situation, the response m^* can easily become so firmly established as to appear inescapable.

Optimal vs. Suboptimal Equilbria

In the symmetric case which we are considering, the noncooperative equilibrium point m^* is defined, formally, by the equation

$$p_1(m^*, m^*, ..., m^*) = \max_m p_1(m, m^*, ..., m^*) , \qquad (*)$$

where the maximum on the right-hand side extends over all possible moves of the first player. This equation describes the equilibrium which will develop as each participant in a symmetric game *privately* adjusts his move to yield a personal optimum, given the moves which others are observed to make. In general this equation selects quite a different point than does the equation

$$p_1(m^*, ..., m^*) = \max_m p_1(m, m, ..., m) , \qquad (+)$$

which describes the *collective* decision that would be taken if all players *first* agree to make the same move and then collectively choose the move which will secure greatest general benefit.

The example games given in the boxes on pp. 125 and 126 serve very well to show that although the m^* *and* m^+ selected by these two quite different equations can accidentally be the same, they can also differ radically. For example, consider the game Stagnation. Here there are three and only three symmetric equilibria: that in which all participants make the move "stay"; that in which all participants make the move "go"; and that in which every participant plays "stay" with probability x and "go" with probability $1 - x$, where x is chosen so as to imply a zero advantage of "stay" over "go" for a player who chooses to deviate from the average. If all players move "go," then each receives $1, the maximum possible payoff in the game, so it is quite unsurprising that this should be an equilibrium. If all players move "stay," then each receives $0, making it somewhat more surprising that this pattern of play should be an equilibrium. Nonetheless it is, since once this pattern of play is reached no player can deviate from it alone without suffering a penalty; in fact he will suffer the penalty unless he can persuade a *majority* of the other players simultaneously to change their moves. Equilibria of this sort, in which it is possible for the situation of *every* player to be improved if the players cooperate suitably, even though it is impossible for a player by himself (and conceivably even small coalitions of players by themselves) to deviate from the equilibrium without suffering, are called *suboptimal equilibria*. We may therefore say that our stagnation game has several equilibria, one optimal and providing the maximum possible payoff $1 to every player, the others suboptimal. Of course, everyone who has fumed through those periods of immovability which affect groups of people as large as half a dozen trying to go together to lunch or to the theater will realize that the suboptimal equilibrium of this model describes a very real phenomenon.

It is even possible for a game to have exclusively suboptimal equilibria. This is clearly shown by our example game Litter (p. 126) which models the dropping of litter in streets. Suppose that all players but one play "drop" with probability x, and "dispose" with probability $1 - x$. Then the payoff to the last player is $-(100/36)-(1000/10)$ if he drops, but is $-5-(999/10)$ if he plays "dispose." Hence every player will "drop," and the equilibrium value of the game for every player will be $-(100/36)-100 = -103$. On the other hand, if the pattern of moves were different, i.e., if every player played "dispose," then the payoff to each player would be -5. This shows that the unique equilibrium of Litter is very distinctly suboptimal.

This result appears paradoxical at first sight, but it has a very simple explanation. The payoff to each player consists of two terms. One, which is either $-(100/36)-(1/10) \cong -3$ or is -5, is directly controlled by his own move; the other term, which in the preceding calculation appeared as $-999/10$, and which in the situation that our model is intended to represent

corresponds to environmental degradation through the accumulation of heaps of litter, is controlled by the moves of others.
Since each player considers the bulk damage represented by the
term 999/1000 to have been wreaked irrespective of his own
tiny contribution to it, his own most advantageous move is to
"drop," which of course subtracts a tiny amount from the payoff
of all other players.

It is instructive to consider what happens if we *decrease* the
payoff function in this model by increasing the fine which must
be paid by persons choosing to "drop" (which they pay in the
case, whose probability is 1/36, that they are caught). Let the
amount of this fine be F; then "drop" remains preferable to
"dispose" as long as $(F/36) + (1/10) < 5$. When F rises above
$5 \times 36 = \$1.80$, "dispose" becomes preferable. For F this large,
the equilibrium payoff to every player is -5. This shows that
the equilibrium value of a game with suboptimal equilbria can
increase (even dramatically) if the game's payoff function is *decreased* in some appropriate fashion. Our calculation also shows
that for minor offenses of this type to be suppressed, the penalty
for the offense, multiplied by the probability of apprehension,
must exceed the gain obtained by committing the offense. Among
other things, this explains why the penalties for littering along
highways (where the probability of apprehension is very low)
must be as large as they typically are.

These game-theoretic arguments suggest a very compelling
objection against classical free-market arguments. The market
equilibria m^* selected by Adams Smith's invisible hand are
generally not the same as the collective maximum m^+. To identify m^* with m^+, simply because their defining equations both
involve maximization, is a logical fallacy comparable to identification of the assertions "every man has a woman for wife"
with "a certain woman is the wife of every man." Our examples
show that the equilibrium return $p_1(m^*, \ldots, m^*)$ to every player
in a symmetric game can be much inferior to the symmetric
maximum return; indeed $p_1(m^*, \ldots, m^*)$ to every player in a
symmetric game can be much inferior to the symmetric maximum return; indeed, $p_1(m^*, \ldots, m^*)$ can lie very close to the

absolute minimum possible value of the payoff function. If a collection of players find themselves trapped in so decidedly suboptimal an equilibrium, they may desparately require some high degree of centralized decision-making as the only way to escape catastrophe. Practical instances of this general fact are known to everyone; hence arise "heroic leaders."

Equilibria in the Economic Sphere

How then can it be argued that natural equilibrium in the economic sphere is generally optimal (whereas, as we have seen, equilibria in more general models can often be suboptimal)? To focus on this question, we must not continue to confine ourselves to the consideration of logically informative but possibly misleading thumbnail examples, as we have done till now, but must go on to give a game-theoretic account of at least the principal forces affecting the overall behavior of a real economy. However before doing so it is irresistibly tempting to use the intellectual tools we have developed to dispose of two persistent catchphrases of popular economic discussion. The first holds that government interference in the economy is necessarily undesirable simply because it is economically unproductive, i.e., "All that a government can do is take money away from some people and give it to others, and charge both for doing so." A second closely related pessimistic phrase is: "There is no such thing as a free lunch."

To contest the first slogan, let us begin by admitting its major premise: that in direct terms the economic activities of the government are purely unprofitable, since in direct terms government can only penalize and not create, so that its activities always wipe out opportunities which free individuals might have been able to exploit; and since to do either some adminstrative cost must be incurred. To go on from this premise to the conclusion that the economic interventions of government cannot be generally and substantially beneficial is to assume that the equilibrium value of a game cannot be increased by reducing its payoff function, an assumption which the preceding examples have shown to be entirely invalid. As to the Free Lunch, our examples also show that by coercively transferring payoffs from some of the players of a game to others, one can (provided that a pattern of transfers is suitably chosen) shift the game's equilibrium in such a way as to leave *every* player with a larger actual payoff than before. In any such situation there is, indeed, such a thing as a free lunch. (It would be wrong, however, for us to conclude merely from the logical existence of such phenomena that they actually occur in the economy as it really is.)

Relevant Economic Facts

To determine if the natural equilibrium of the economy is optimal or suboptimal we must introduce additional empirical material, painting a picture of the dominant forces in the real economy detailed enough for us to derive at least a coarse idea of its motion. Unfortunately, to do so we will have to step onto considerably thinner ice than that which has borne us till now, since the economy is an enormously varied and complex thing, valid generalizations concerning which are hard to come by. In forming the picture which follows we are in effect selecting one particular model out of many possible mathematical models of reality; to judge the adequacy of this model fairly requires empirical information going far beyond the boundaries of the present article. It is entirely possible that by selecting another (perhaps better) model we might be led to conclusions diametrically different from those which we will draw. These hazards notwithstanding, we shall identify the following as describing the principal forces at work in the economy:

Material profitability of production. Let the various goods produced within an economy be numbered from 1 to k. To produce one unit of the jth commodity some (nonnegative, but possibly zero) amount π_{ji} of the ith commodity will have to be consumed as "raw material," and in addition some quantity ϕ_{ji} of the ith commodity will have to be used as "capital equipment" and this will be tied up temporarily during the actual production of the jth commodity. If the economy is not doomed to progressive exhaustion, there must exist some overall arrangement of production which is physically profitable, i.e., an arrangement which produces an amount v_i of every commodity which exceeds the amount that is consumed in production.

Pattern of distribution. Suppose that we designate the amount of labor (measured, e.g., in man-hours) required for the production of one unit of the jth commodity as π_{j0}. Then one part of the net physical economic product (this physical net being $v_i - \Sigma_{j=1}^{k} v_j \pi_{ji}$) is distributed as *wages*; the portion that then remains may be called national physical net *profit*, and is available either for consumption or for addition to the national stock of capital or consumable supplies. Wages are distributed in proportion to hours worked (i.e., to $\Sigma_{j=1}^{k} v_j \pi_{j0}$); each individual has the right to claim part of the national physical net profit (as a dividend) in proportion to the portion of capital which he owns.

Pattern of consumption. Of the portion of the net national product available to them, individuals (or families) will elect to consume a part, but will also wish to set a part aside for future consumption. The part set aside may be intended as a retirement fund, an accumulation for planned major expenditure, a reserve for contingencies, or a bequest to family or society. Generally speaking, the proportions of income reserved from immediate consumption will rise with rising income (for various reasons: e.g., people will not want their retirement income to fall drastically short of the income to which they have grown accustomed). At sufficiently high income levels the desire to consume will begin to saturate, and the bulk of income above such levels will be reserved for investment. On the other hand, individuals not constrained to do so will not react to a fall in income by immediately reducing their level of consumption; instead they will at first maintain their level of consumption by consuming a portion of the wealth to which they have claim, cutting back their habitual level of consumption only gradually.

Income reserved from consumption takes the form of claims to a portion of the inventory, capital plant, and income of productive enterprises; also claims on the future income of enterprises or of other individuals; and also exemptions from anticipated future tax payments (this latter being "government debt").

The profit maximization strategy of productive enterprises. A firm manufacturing some given commodity maintains an inventory, selling off portions of this inventory as orders are received, and initiating new production to rebuild inventory as appropriate. In attempting to maximize a firm's profitability, its managers manipulate the three control variables available to them: product price, level of current production, level of investment expenditures. If the production of one commodity consistently remains more profitable than that of another, investment will be directed towards production of the more profitable commodity. Thus over the long term we must expect the rates of profit on all types of production to converge to a common value ρ. When this "investment equillibrium" is reached, the price of each commodity will be the price which yields the normal rate of profit ρ on its production. (Thus, if the wage rate per man-hour is w and p_j is the price of the jth commodity, we expect the managers of all firms to be constrained in the long run by the following equation in their choice of prices:

$$p_j = \sum_{i=1}^{n} \pi_{ji} p_i = \pi_{j0} w + \rho \sum_{i=1}^{n} \phi_{ji} p_i .)$$

A different and shorter-term set of considerations enters into the determination of current production levels. The commodities that we have been considering, which may for example be clothing, automobiles, children's toys, or residential dwellings, are not as homogeneous or time-invariant as our argument till now may have suggested. Indeed, as time goes along the sale value of such items is progressively undermined by changing factors of fashion and technological advance, by seasonal considerations, and by storage and upkeep costs applying even to unused commodities. For example, clothing not sold promptly will miss its season and have to be held for a year, by which time it may well have grown unfashionable and have to be sold at a drastic markdown; similarly, computers built and not sold will rapidly become valueless as technology advances. To model the effect of all the important technological and deterioration effects of this kind, we can oversimplify and assume that all commodities are slightly perishable, i.e., that any part of a given commodity stock which is neither consumed nor used as raw material during a given nominal cycle of production decays at some specified rate. This decay rate can be quite large, and if it is assumed to be larger than the physical profitability of production it is obvious that all firms will have to tailor planned inventories rather closely to expected sales. This basic fact traps all firms in a situation resembling the Stagnation game discussed above: all firms collectively will profit as the level of economic activity rises, but no firm individually can afford to let its own level of production outrun that of all the others. In this relationship, which as we have seen can lead very directly to a stable sub-optimal equilibrium, we see a root cause both of recessions and of protracted periods of economic stagnation.

Factors related to the housing and mortgage market. In purchasing a home, few people pay cash out of pocket. Instead, they require mortgage financing. The appeal, and even the possibility, of purchasing a house depends critically on the current state of the mortgage market. Two related but significantly different factors, namely mortgage rates and mortgage terms, must be considered. High interest rates are seen by the prospective house purchaser as high monthly payments, in direct linear proportion. Thus rising interest rates choke off real estate sales. Unfavorable mortgage terms can erect an even more impenetrable barrier against home purchases. In periods of high economic activity, commercial demand for short-term loans will bring banks close to the limit of their lending ability. A bank finding itself in this situation can afford to be choosier about the mortgage loans it makes; in particular, it can afford to insure the quality of its mortgages by increasing the proportion of total purchase price that the prospective purchaser is required to pay out immediately. The prospective home purchaser will see this as an intimidating increase in required down payment, e.g., the amount of cash that he must come up with to purchase a $60,000 home may rise from 15%, i.e., $9,000, to 25%, i.e., $15,000. Those unable to meet the rise in terms must quietly retreat from the would-be purchase.

Stable spending patterns of retired persons and government. If we ignore the small minority of persons who derive a substantial portion of their income from either dividends or royalties, we can say that the income of nonretired persons is proportional to the number of hours that they work, and hence is controlled in close to linear fashion by the general level of economic activity. (However, unemployment insurance payments by the government serve to lessen the impact on income of a diminished level of economic activity.) But the income of retired persons, and hence also their spending, is constant (in cash terms, if they are living on annuities or other fixed-dollar pensions, or in real terms, if they support themselves by selling previously accumulated stocks or real estate), and hence their spending levels (in cash or real terms, as appropriate) are also constant. Much the same remark applies to the activities of government, which continue at legislated levels in substantial independence from the oscillating level of economic activity which surrounds them. (Of course, the federal government has a considerably greater ability to legislate a given level of government demand, independent of the level of activity in the remainder of the economy, than do the state and local governments, since any state or municipality whose tax rates remain above average for long will soon find its share of economic activity, and hence its tax base, shrinking as economic institutions able to do so flee. We may say here that the overall taxing power of the separate states is held in check by the fact that they are independent players in a fifty-sided game of "offer better tax terms.")

Econometric Models

We have now drawn a complete enough sketch of the principal determinants of economic activity to explain the structure of econometric models of the sort actually used to track the economy and to guide national economic authorities. Of course, since the quantitative material, motivational, and planning factors which determine the motion of the real economy are enormously varied, the development of a quantitatively reliable model of the economy is not easy and the best currently available models are far from perfect. Nevertheless, carefully constructed linear econometric models generally give good forecasts within a fore-casting range of about a year. Beyond a one-year range, the economic behavior predicted by such models will differ pro-gressively from real events owing to the accumulating effects of inaccuracies in the model and nonlinearities in the real economy. However, a year's advance prediction gives ample time for policies to be adjusted, if only the will to do so is pre-sent and is buttressed by a general qualitative understanding of what is happening.

In constructing such models, one begins with a list of major influences on economic activity like that which we have pre-sented above, but expands the crude account which we have given into a more detailed but still informal picture of the manner in which econometrically critical consumer and corporate plan-ning variables (e.g., purchasing plans, production, inventory, and investment levels) are likely to depend on objective factors signficant to individual planners (e.g., income, security of income, sales levels, rate of inventory turnover, plant capacity in use, long- and short-term interest rates). This informal reason-ing is used to set up a collection of equations, generally of the linear form

$$n = c + d_1 x_1 + \cdots + d_n x_n ,$$

in which the terms which can occur are prespecified, but in which the numerical coefficients $c, d_1, \ldots, d_n$ are initially undetermined. (A specific example of such an equation — the Wharton equation used to predict levels of residential housing construction — is described in the box on p. 145.) However, it must be stated that the process by which econometric equations are formed is still uncomfortably subjective, since existing techniques generally do not give us any clear way of focusing objective data strongly enough to discriminate sharply among alternative choices of terms to include in a model.

Nevertheless, in spite of the substantial quantitative and even qualitative uncertainty which adheres to them, econometric models do embody, and their success does tend to confirm, the general picture of economic forces. In particular, detailed examination of these models does tend to support our earlier suggestion that the totally unregulated economy can behave suboptimally, and hence suggest that recessions can be understood as suboptimal equilibria of the stagnation type, whose operative mechanism is found in the fact that no firm can afford to let its own level of production outrun that of all the others. From this point of view, the most noteworthy mechanism of the familiar cycle of boom and recession seems to be the gradual accumulation of inventories during boom periods, and the manner in which the rising interest rates associated with such periods act to choke off housing sales. Once these forces have caused the economy to fluctuate downward, automobile sales will ordinarily respond sharply, and much of the dynamic oscillation associated with a recession will tend to be concentrated in the automobile sector and related industries.

To remove this suboptimality, economic policy-makers have attempted since the time of Keynes to shift the economic equilibrium toward higher levels of production by raising the total of personal, investment, and government demand that determines the level of production which can be sustained without inventories beginning to accumulate to an unreasonable degree. Policies of this general kind are administered in various ways, for instance, by lowering personal income taxes, by increasing pension payments, by increasing the number of pensioners or the size of the armed forces, or by encouraging private and public investment through investment-related rebates and guarantees to private firms or through major direct public investments in, e.g., the road system, municipal improvements (including housing), or the national armament. It deserves to be noted in connection with this comment on anti-recession policy that the policy-maker's eye must focus primarily on the possibility of continuing economic underperformance rather than on the ups and downs of the business cycle, just as the prudent car owner ought to be more concerned with poor engine performance (whose cost will mount mile after mile) than with occasional, even if alarming, engine knocking. We note in this connection that the most destructive aspect of the business cycle

---

Residential Housing Construction in the Wharton Econometric Model

The 1967 Wharton econometric model, which served as a prototype for many subsequent econometric models, contains an equation used to fit the observed behavior of investment in residential housing. The assumed form of this equation reflects the general factors (such as total disposable income, mortgage availability), thought to influence residential housing purchases. The precise form of this equation, which we present as an example in order to make the structure of such models vivid, is:

$$I_k = 58.26 + 0.0249\,Y - 45.52 \left(\frac{p_h}{p_r} \right)_{-3}$$

$$+ 1.433(i_L - i_s)_{-3} + 0.0851(I_h^s)_{-1} \; .$$

Here I_k is the rate of investment in residential housing in billions of dollars, measured in a given quarterly (3 month) period; Y is total disposable personal income (in billions of dollars) for the same period; p_h is an indicator of the average price of housing and p_r an indicator of average rent levels. (The subscript -3 attached to the fraction p_h/p_r indicates that the ratio to be used is that calculated for a period three quarters (i.e., 9 months) prior to the period in which I_k is measured.) The terms i_L and i_s are the long- and short-term interest rates (again taken from a period 9 months previous to the measurement of I_h), and I_h^s is the rate of housing starts (measured three months, i.e., one quarter, prior to the measurement of I_h).

---

may be the way in which it confuses macroeconomic policy, since the cyclic oscillation of the economy insures that if one applies almost any measure, sufficient or insufficient, and then waits long enough, an encouraging boomlet of economic improvement will allow one to argue for the correctness of one's policy.

The above appreciation of the mechanism of recessions, together with the general game-theoretic reasoning developed in the preceding pages, suggests an attempt to analyze other significant economic problems in the same way as suboptimal equilibria of game-like models, and to see what cures recommend themselves from this viewpoint. To follow this thought, we begin by noting that the anti-recession measures sketched above act to dissolve suboptimal equilibria by rewarding players who avoid available but globally undesirable courses of action, e.g., by subsidizing investment that would otherwise seem unwise or consumption that would otherwise be impossible. But these measures are not entirely unproblematical, since they can push the economy toward suboptimalities of other sorts. Note in particular that to implement policies like those we have listed, a government needs to dispose of some form of reward, which must be available readily and in massive quantities. The items of reward used in practice are notes of government indebtedness of various kinds; the sum of all these notes constitutes the national debt. These notes are most rationally viewed as certificates of exemption from future tax payments (since the notes themselves can be presented in payment of taxes). This fact gives the notes their basic value; and the inducement to hold them can be increased as much as desired by making them interest-bearing at a suitable rate. At first sight, it might appear that this debt-generating technique cannot safely be used over the long term, since one might fear that the accumulation of larger and larger masses of exemption certificates would eventually undermine the taxing ability which gives these certificates their value, leading ultimately to a collapse in the value of government notes and of money as well, that is, to a hyperinflationary disaster like the German hyperinflation of the 1920s or the Chinese hyperinflation of the 1940s, which would constitute another kind of suboptimal collective phenomenon. To eliminate this phenomenon, a monetary authority must ensure that whenever the accumulation of these certificates becomes troublesome some part of them will be drawn off. This can be done simply by imposing some appropriate form of property tax. In fact, such a tax can be made to act continuously, for example, by setting the estate-tax level appropriately.

Thus from our game-theoretic point of view we can recognize inflation itself as a price-related suboptimality of the economy. Analysis of theoretical models suggests (and empirical econometrics seems to confirm) that in the economy as it is, the general level of prices and wages is not determined absolutely (e.g., as an equilibrium level in some macroeconomic price/wage model), but simply floats about under the control of purely frictional forces. The Wharton model equations, for example, state (see box on p. 148) that prices follow wages and wages follow prices, and hence suggest that the overall wage/price level does not tend to any equilibrium. Moreover, these equations indicate that the frictional force which holds back rises in the general wage level depends directly on an exceedingly undesirable economic phenomenon, namely, unemployment. This being the case, we need to consider what would be the least unacceptable response to a coupled inflation-unemployment suboptimum.

U.S. macroeconomic policy has been considerably disoriented during the past half decade by confusion about how to deal with this problem. The significant facts reflected in the models we have discussed suggest that there may exist no satisfactory response to it other than to admit some degree of government intervention in the formation of prices and wages. At the same time, examination of the relatively mild consequences of moderate levels of inflation suggests that a mild frictional intervention, which deliberately tolerates some modest level of inflation, thereby avoiding the rigidities of a more strict wage/price control system, may be sufficient. If, for example, a central authority counts on a 2% annual rise in productivity and is prepared to tolerate a 4% inflation rate, wage/price control might consist in a law providing that any salary rise of more than 6% per annum would be taxed (and even withheld) at a close to 100% rate, and that a corporation whose prices rose by more than 5% in a given year would be liable for a surtax on the resulting excess revenue.

Of course, one must always fear that coarse, rigid centralized economic interventions will hamper the private sector's stunning ability to discern desired products, services, and productive opportunities in general. Nevertheless, the game-theoretic point of view central to the present article shows that these objections do not necessarily apply to interventions which, by grasping and

Prices and Wages in the Wharton Econometric Model

In the Wharton econometric model, the general price level p_m of manufactured goods is determined by the equation

$$p_m = -0.170 + 0.514(W/X) + 0.2465(X/X_{\max})$$

$$+ 0.6064 \left(\frac{(p_m)_{-1} + (p_m)_{-2} + (p_m)_{-3} + (p_m)_{-4}}{4} \right).$$

Here W/X is the labor cost per unit of product, X the overall level of manufacturing production, and $X_{\max}$ the estimated maximum capacity level of production. Translated into heuristic terms, this equation simply states that prices are proportional to labor costs, but will be raised somewhat above this as firms find themselves reaching capacity production; however, (as is shown by the last group of terms) prices do not adjust immediately to changed costs or to pressure on capacity, but adjust only gradually over a period of something like one year. The Wharton model's equation for the wage level (of manufacturing employees) is

$$W = W_{-4} + 0.050 + 4.824(p_{-1} - p_{-4}) - 0.1946(W_{-4} - W_{-6})$$

$$+ 0.1481 \left(\frac{(U - U^*)_{-1} + (U - U^*)_{-2} + (U - U^*)_{-3} + (U - U^*)_{-4}}{4} \right)$$

Here, W designates the wage level, p the price level, U the general rate of unemployment, and U^* the rate of unemployment among males 25-34 years of age, so that the difference $U - U^*$ measures the extent to which the pool of first-hired, last-fired employees is fully used. In heuristic terms, this equation shows the wage level rising five percent per year, but rising additionally in proportion to the rise in prices during the preceding year, and also rising somewhat less than otherwise expected if a substantial wage rise was registered the year before (since poor raises last year argue for better-than-average raises this year). Moreover, unemployment in the first-hired, last-fired pool acts to restrain the rate at which wages rise. (However, this rate of rise is resistant to other categories of unemployment.)

manipulating some small and well-chosen set of macroeconomic parameters, move the economy away from an undesirable equilibrium. Indeed, a policy of this kind should be regarded not as contradicting but as perfecting the approach advocated by Adam Smith, specifically by adjusting the global economic environment in a way calculated to ensure that his "invisible hand" will really lead the participants in an economy to an optimal equilibrium, i.e., to a situation in which no further benefit without a compensating sacrifice is possible. It is, of course, true that even interventions which merely block off options that lead directly to a suboptimal situation will be perceived by some, and perhaps by many, of the participants in an economy as arbitrary restrictions which cut off profitable and desirable alternatives. But this view can be mistaken, since the alternatives being cut off are the gateway to an overall process, irresistible by any single firm or limited group of firms, in which everyone's most rational efforts lead all to a less desirable situation than all might otherwise have enjoyed.

1978

References

General

1. DAVIS, MORTON D. *Game Theory, A Non-Technical Introduction.* Basic Books, New York, 1970.
2. DRESCHER, MARVIN. *Games of Strategy — Theory and Applications.* Prentice-Hall, Englewood Cliffs, 1961.

Technical

1. BACHRACH, MICHAEL. *Economics and the Theory of Games.* Macmillan, London, 1976.
2. DUSENBERRY, J. S. FROMM, G. KLEIN, L.R. and KUH, E. (Eds). *The Brookings Quarterly Econometric Model of the United States.* Rand-McNally, Chicago, 1965.

The theory and technique of econometric model construction.

1. KLEIN, LAWRENCE R. and EVANS, MICHAEL K. *The Wharton Econometric Forecasting Model.* Economics Research Unit, Department of Economics, Wharton School of Finance and Commerce, Philadelphia, 1967.
2. VON NEUMANN, JOHN and MORGENSTERN, OSCAR. *Theory of Games and Economic Behavior.* Princeton University Pr, Princeton, 1955.

II

CHAPTER ELEVEN

Complicating Mathematics

AT THE TURN OF THE CENTURY, the Swiss historian Jakob Burckhardt, who, unlike most historians, was fond of guessing the future, once confided to his friend Friedrich Nietzsche the prediction that the Twentieth Century would be "the age of oversimplification."

Burckhardt's prediction has proved frighteningly accurate. Dictators and demagogues of all colors have captured the trust of the masses by promising a life of bread and bliss, to come right after the war to end all wars. Philosophers have proposed daring reductions of the complexity of existence to the mechanics of elastic billiard balls; others, more sophisticated, have held that life is language, and that language is in turn nothing but strings of marble-like units held together by the the catchy connectives of Fregean logic. Artists who dished out in all seriousness checkerboard patterns in red, white, and blue are now fetching the highest bids at Sotheby's. The use of such words as "mechanically," "automatically" and "immediately" is now accepted by the wizards of Madison Avenue as the first law of advertising.

Not even the best minds of Science have been immune to the lure of oversimplification. Physics has been driven by the search for one, only one law which one day, just around the corner, will unify all forces: gravitation and electricity and strong and weak interactions and what not. Biologists are now mesmerized by the prospect that the secret of life may be gleaned from a double helix dotted with large molecules. Psychologists have prescribed in turn sexual release, wonder drugs and primal screams as the cure for common depression, while preachers would counter with the less expensive offer to join the hosannah-ing chorus of the born-again.

It goes to the credit of mathematicians to have been the slowest to join this movement. Mathematics, like theology and all free creations of the Mind, obeys the inexorable laws of the imaginary, and the Pollyannas of the day are of little help in establishing the truth of a conjecture. One may pay lip service to Descartes and Grothendieck when they wish that geometry be reduced to algebra, or to Russell and Gentzen when they command that mathematics become logic, but we know that some mathematicians are more endowed with the talent of drawing pictures, others with that of juggling symbols and yet others with the ability of picking the flaw in an argument.

Nonetheless, some mathematicians have given in to the simplistics of our day and when it comes to the understanding of the nature of their activity and of the standing of mathematics in the world at large. With good reason, nobody likes to be told what he is really doing or to have his intimate working habits analyzed and written up. What might Senator Proxmire say if he were to set his eyes upon such an account? It might be more rewarding to slip into the Senator's hands the textbook for Philosophy of Science 301, where the author, an ambitious young member of the Philosophy Department, depicts with impeccable clarity the ideal mathematician ideally working in an ideal world.

We often hear that mathematics consists mainly in "proving theorems." Is a writer's job mainly that of "writing sentences"? A mathematician's work is mostly a tangle of guesswork, analogy, wishful thinking and frustration, and proof, far from being the core of discovery, is more often than not a way of making sure that our minds are not playing tricks. Few people, if any, had dared write this out loud before Davis and Hersh.[1] Theorems

---

[1] The Mathematical Experience, Birkhäuser-Boston, 1981.

are not to mathematics what successful courses are to a meal. The nutritional analogy is misleading. To master mathematics is to master an intangible view, it is to acquire the skill of the virtuoso who cannot pin his performance on criteria. The theorems of geometry are not related to the field of Geometry as elements are to a set. The relationship is more subtle, and Davis and Hersh give a rare honest description of this relationship.

After Davis and Hersh, it will be hard to uphold the *Glasperlenspiel* view of mathematics. The mystery of mathematics, in the authors' amply documented account, is that conclusions originating in the play of the mind do find striking practical applications. Davis and Hersh have chosen to describe the mystery rather than explain it away.

Making mathematics accessible to the educated layman, while keeping high scientific standards, has always been considered a treacherous navigation between the Scylla of professional contempt and Charybdis of public misunderstanding. Davis and Hersh have sailed across the Strait under full sail. They have opened a discussion of the mathematical experience that is inevitable for survival. Watching from the stern of their ship, we breathe a sigh of relief as the vortex of oversimplification recedes into the distance.

1981

CHAPTER TWELVE

Mathematics and Its History

THE HISTORY OF MATHEMATICS is an enticing but neglected field.

One reason for this situation lies in the nature of intellectual history. For any theoretical subject x, telling the story of x is not a conceptually distinct undertaking from describing the theory of x, though the two presentations often appear in different guises. The readership of a serious history of x will thus be largely limited to the few specialists in x, a small circulation at best. Worse yet, mathematics, or science for that matter, does not admit a history in the same sense as philosophy or literature do. An obsolete piece of mathematics is dead to all but the collector of relics. Discovering that the Babylonians knew harmonic analysis may be an astonishing feat of scholarship, but it is a supremely irrelevant piece of information to working scientists. Few of the serious historians of mathematics have realized this; as a result, we are saddled today with competent histories of Greek and Renaissance mathematics, but we sadly lack such items of burning interest as "The Golden Days of Set Theory (1930–1965)," "Topology in the Age of Lefschetz (1924–1953)," "The Beginnings of Probability (1932–···)," to cite but a few possible titles.

Faced with these and many other problems, Morris Kline[1] has chosen the courageous avenue of compromise. In his book,

[1]Morris Kline, *Mathematical Thought from Ancient to Modern Times*. Oxford University Press, NY, 1972.

the Greeks get a 15% cut, the Egyptians are whisked on and offstage, the Arabs and Renaissance together make a fleeting 10% appearance, and the drama begins with René Descartes on Chapter 15 out of 51.

Synthetic geometry is Morris Kline's first love, and he returns to it with predictable regularity. It is thus no wonder that his treatment of Greek geometry succeeds in being readable, at times downright interesting. Ancient mathematics has a way of appearing to us coated with a dusty layer of weirdness, which historians have occasionally attempted to brush off with a phony emphasis on the human angle, as if we could reconstruct what made the Greeks tick without the aid of science fiction. Kline's solution, instead, is to translate into contemporary language and values the best of Greek geometry, with brief mention of the authors. His is a history of ideas, not one of men. Nonetheless, one is pleased to find throughout the book a scrupulous fairness in giving the little guy his due, and in emphasizing the greatness of some of the lesser known names. For example, Apollonius gets equal billing with Euclid, as he should. To pick another example at random, in the farewell chapter to synthetic geometry, the names of Pieri and Veronese are mentioned together with that of Hilbert, though not on a par; but then, Hilbert had the glory of Göttingen behind him, while Pieri — whose treatment of the foundations of geometry Russell considered superior — was sweating at Parma U. The book abounds with such touching episodes, though a really thorough straightening of mathematical priorities is not carried out; it might have made quite a shocker.

The climax comes in the chapters on nineteenth century analysis. Kline's expertise in the subject makes his presentation a delight. His device of combining historical development with a clear, linear description of the material turn the book into a useful *précis* of mathematics up to the turn of the century. The lucid chapter on asymptotic series — inexplicably combined with the obsolete notion of summability — makes one wonder why such an obviously fundamental topic has not found its way into the standard calculus curriculum. Similarly, his two-page presentation of Fredholm's ideas on integral equations leaves one filled with wonder. On the other hand, the two chapters on differential geometry might have gained by the use of contemporary notation; in their present form, they will be all but inaccessible to students trained in the last fifteen years. But Kline has a bone to pick with certain contemporary mathematical schools. One may wonder why the chapters on functional analysis and topology are also presented with a decidedly conservative slant, which is at variance with present-day practice in graduate courses, and does not relate to the current problematic of these subjects. Algebra and logic, the two triumphant if not always amicable contenders for center stage in present-day pure mathematics, get only a fleeting mention. There is not a word on group representations, classical as the subject is, and the deeper results in number theory are hardly touched upon, perhaps because these subjects call for higher mathematical sophistication, and Kline wants to keep his requirements down to the second-year calculus level.

Morris Kline's abrupt stop at the nineteenth century, as well as his recurring emphasis on synthetic geometry, are not simply a whim. This is not only a handbook of basic mathematical culture, nor merely a good reference book of mathematical history. It is also meant as a message; it stresses and at times betrays a forceful position on some controversial issues in contemporary mathematics. A grasp of these issues is a prerequisite for the understanding of the book's unity of purpose. Briefly, these issues are: (1) the originality of contemporary mathematics, and (2) the raging battle over the "New Math" reform in the high schools.

The first accounts for the stress on the remote past. It is the author's contention that contemporary mathematics runs a poor second to the mathematics of the past century. The excesses of abstraction and notation of the present conceal a dearth of ideas and lack of novelty, or so the author claims. For Kline, mathematics is primarily a body of facts; proof is only the tool for their confirmation. Some contemporary mathematicians think otherwise: for them, according to Kline, the results of mathematics are contingent upon and secondary to the formal-deductive structure of mathematical reasoning.

There are strong points for both sides. Kline's book is in part an implicit argument in support of the author's thesis. It is a showcase display of the great facts of mathematics, presented in their shining light and eternal value. On leafing through the book, one is struck how hard indeed it is to find in our century new mathematical theories as obviously useful as the calculus of variations or the oscillation theory for solutions of differential equations. (Strangely, in physics the opposite seems to be true.) However, Kline's assertion that "abstract algebra has subverted its own role in mathematics" is at best a partisan one. If we are to admire Descartes for his reduction of geometry to algebra, then why is contemporary (commutative) algebra not equally admirable for successfully unifying and cross-fertilizing such disparate undertakings as number theory, algebraic geometry, differential geometry and topology? Present-day mathematics is poorer in old-fashioned facts and more abundant in concepts, but its avowed aim is a synthesis and simplification which may bring about far more sweeping changes than a simple accumulation of facts would. Ours is the age of "terrible simplifiers" forecast by Jakob Burckhardt, whose effects upon the future may still be in the realm of conjecture, but will not in any case be negligible.

On the issue of the "New Math," Morris Kline is a paladin of the Old Order. By wiping synthetic geometry out of the curriculum and replacing it with subjects that make fewer demands for the student's active participation, this reform is threatening to turn math from a "hard" into a "soft" subject. The old emphasis on math as a collection of facts about the world has been replaced by a stress on deductive argument and formal thought. An increasing number of mathematicians are coming to doubt the value of this reform; Kline's book, intended to show teachers and other practitioners "what good math used to be like," will be a rallying point for the opposition.

It is easy to find something to criticize in a treatise 1200 pages long and packed with information. But whatever we say for or against it, we had better treasure this book on our shelf, for as far as mathematical history goes, it is the best we have.

1974

CHAPTER THIRTEEN

Academic Responsibility

ACADEMIC COMMUNITY MEANS UNIVERSITIES, and we are faced at once with a difficulty, for it is no longer certain that a consensus within the academic community of today can be achieved as to what a University is and what it should do.

The unrest of the late sixties and early seventies left the academic community divided on almost every issue and caused it to be preoccupied with matters much more immediate and far less abstract than the ones we are discussing here.

That part of the academic community, and it is far from being a negligible minority, which believes that the University should become an instrument of immediate social change is not likely to become involved in a *dispassionate* explanation of *long-range* effects of what we do now in science and technology.

The words *dispassionate* and *long-range* are italicized because in my perhaps old-fashioned view they are inexorably tied to the social responsibility of the academic community.

Universities by an unwritten covenant with Society evolved to be largely concerned with problems on a time scale of a generation or longer.

The order of magnitude of this time scale was impressed upon me while still a student in Poland. In 1935, or was it 1936, encouraged by the rise of Hitlerism in Germany, ultra nationalistic and reactionary students forced the Minister of Education to decree the infamous "ghetto benches," forcing Jewish students to sit on the left side of each lecture room.

Since Universities were, on paper at least, autonomous institutions, the Minister's order had to be countersigned by the Rector of each University.

All but one signed with hasty docility. The sole exception was the Rector of the University of Lwow, which happened to be my University.

A noted plant scientist and a man of great integrity and moral courage, Professor Kulczynski resigned in protest and in a letter to the Minister said that, "if one destroys a power plant, it is dark at once, but if one destroys the Universities, it is dark fifty years hence."

The story as such has perhaps no direct bearing on matters we are discussing, but the estimate of the time scale is as valid now as it was under wholly different circumstances for entirely different reasons 40 years ago.

It is, in fact, the length of the time scale that produced the stereotype of the University Professor as the impractical dreamer who in a self-imposed exile in an Ivory Tower has neither the appreciation nor the understanding of the "real problems" of the "real world."

All this changed, of course, when some "impractical dreamers" gave the world atomic and nuclear weapons and, in spite of the indifference and even hostility of "practical" tycoons, became instrumental in the creation of a multibillion dollar computer industry.

But, having tasted sin, to paraphrase the late Robert Oppenheimer's familiar expression, the scientists of Academe have also succumbed to the lures of short-term advantages and rewards.

Having helped to create a technology of unprecedented sophistication, science could now use its fruits to increase its own tempo and pace to a level which surpasses the imagination.

Molecular biology is not much more than 30 years old, but already it has reached a stage when some of its most prominent leaders are advocating a halt to certain kinds of experimentation for fear that one may create living things whose effect on our planet, including *Homo sapiens*, might be devastating.

The huge and impressive high energy machines make discoveries at a rate that renders understanding and assimilation of them so difficult as to border on the impossible.

And, last but perhaps not necessarily least, the computer, which again only 30 years ago was but a figment of the imagination of Alan Turing and John von Neumann, is now so fast and so powerful that it can in a matter of minutes, or at most hours, process models which, imperfect as they may be, predict the economic future of the world for several decades.

All this is a far cry from the leisurely and reflective time scale to which Universities have been geared for centuries.

Our main problem is what may be called the conflict of time scales, and the resolution of this conflict is where our social responsibility primarily lies.

Society on whose support universities depend has no real feeling for the long time scale. It demands immediate results and becomes impatient if they are not forthcoming.

It wants a cure for cancer, not a deeper understanding of the molecular basis of biology; atomic power plants, not new particles; relief from inflation, not abstruse non-linear economic models which predict disaster in some incomprehensibly distant future.

The man on the street who represents Society is even apt to react with hostility when we try to deprive him of today's comforts for the sake of avoiding future dangers which to him are as incomprehensible as they are remote.

The dilemma this poses is clear.

On the one hand, deeper understanding without which there can be no real progress can only be achieved by a slow, difficult and basically lonely process. On the other, there is the constant pressure for getting tangible results long before their implications are fully explored or even understood.

The conflict of time scales is deepened by the accomplishments of technology and the uncritical belief in uncontrolled growth.

The great mathematician Euler complained that on occasions his pencil moved faster than his mind. We are confronted with something similar but on a much grander scale now with potential dangers which stagger the imagination.

The resolution of the conflict of time scales is the most vital social responsibility of the academic community. By tradition and partly by default it is the sole guardian of the long-range commitment. Indispensable as we may find the "technological pencil," we cannot allow it to outrun its mind. And it must sow seeds of future progress while at the same time help eliminate the weeds of the past.

1975

CHAPTER FOURTEEN

Husserl and the Reform of Logic

AN UNBRIDLED AND PASSIONATE INTEREST in foundations has often been singled out as a characteristic trait of both philosophy and science in this century. Nowhere has this trend been more rampant than in mathematics. Yet, foundational studies, in spite of an auspicious beginning at the turn of the century, followed by unrelenting efforts, far from achieving their purported goal, found themselves attracted into the whirl of mathematical activity, and are now enjoying full voting rights in the mathematical senate. As mathematical logic becomes ever more central within mathematics, its contributions to the philosophical understanding of foundations wane to the point of irrelevance. Worse yet, the feverish technical advances in logic in the last ten years have dashed all hope of founding mathematics upon the notion of *set*, which had become the primary mathematical concept since Cantor. Equally substantial progress in the fields of algebra and algebraic geometry[1] has further contributed to cast a shadow on this notion. At the other end of the mathematical spectrum,

---

[1] The fascinating story of the evolution of the notion of set in modern algebra remains to be told, perhaps because it is far from concluded. It began in the thirties, when the discoveries of the nineteenth-century geometers were subjected to rigorous foundation with the help of the newly developed algebraic methods. By way of example, the notion of point, which once seemed so obvious, has now ramified into several different concepts (geometric point, algebraic point, etc.) The theory of categories offers at present the most serious challenge to set theory.

the inadequacy of naive set theory had been realized by von Neumann[2] since the beginning of quantum theory, and to this day the physicist's most important method of research remains devoid of adequate foundation, be it mathematical, logical, or philosophical.

It is a boon to the phenomenologist that the notion of set, together with various other kindred notations, should have become problematic at this time (as predicted long ago by Husserl); thanks to the present crisis of foundations, we are allowed a rare opportunity to observe fundamental scientific concepts in the detached state technically known as "bracketing." As often happens, the events themselves are forcing upon us a phenomenological reduction.

In high-energy physics, an even worse crisis of foundations is looming, and yielding a substantive crop of newly problematic concepts: matter, distance, measurement — to name only a sample — are losing their former evidence and daily gaining in obscurity, while at the same time a mass of unexplained and ever more precise experimental data is waiting for some kind of conceptual explanation. The focus of the crisis seems to be the concept of *time*, heretofore taken for granted and never genetically analyzed by anyone in the sciences. The need for a reform of this concept has now been recognized by scientists of all professions, while the contributions made by phenomenologists to its understanding are widely ignored.

Can Husserl's philosophy and method help us out of these and other foundational predicaments? It is our contention that the way of phenomenology is inevitable in the further development of the sciences. I shall sketch a possible beginning of such an enterprise, by an admittedly inadequate and schematic presentation.

---

[2] Most of von Neumann's work in pure mathematics (rings of operators, continuous geometries, matrices of high finite order) is concerned with the problem of finding a suitable alternative to Boolean algebra, compatible with the uncertainty principle, upon which to found quantum theory. Nevertheless, the mystery remains, and von Neumann could not conceal in his later years a feeling of failure over this aspect of his scientific work (personal communication from S.M. Ulam).

One of Husserl's basic and most firmly asserted themes is — as his readers know all too well — that of the autonomous onto-logical standing of distinct eidetic sciences. Physical objects (such as chairs, tables, starts, and so forth) have the same "degree" of reality as ideal objects (such as prices, poems, values, emo-tions, Riemann surfaces, subatomic particles, and so forth). Nevertheless, the naive prejudice that physical objects are some-how more "real" than ideal objects[3] remains one of the most deeply rooted in Western culture, and within the West, most firmly entrenched in the English-speaking world (most of Hus-serl's critique of Hume pivots around this one issue). A conse-quence of this belief — which until recently was not even per-ceived as such — is that our logic is patterned exclusively upon the structure of physical objects. Mathematical logic has done us a service by bringing this pattern to the fore in unmistakable clarity: the basic noema is the set, and all relations between sets are defined in terms of two of them: $a \subseteq b$ (a is contained in b) and $a \in b$ (a is a member of b).[4] The resulting structure was deemed sufficient for the needs of the sciences up to a few years ago.

The present crisis forces us into a drastic revision of this logic, along the lines that Husserl drew. This revision is, I believe, the unifying intentional theme throughout *The Crisis of European Sciences*.

How is such a revision to be carried out? Gleaning from Husserl's writings, we propose the following preliminary steps.

1. One must first come to a vital, actual realization that the physical object is no longer to be taken as the standard of reality. The techniques of phenomenological and existential description have been developed for just this purpose: while at one end they serve to bracket the physical world and thereby reveal its *con-tingency*, at the other end they bring out the experiential reality

---

[3] Perhaps the clearest exposition and criticism of this attitude is to be found in Nicolai Hartmann's *Zur Grundlegung der Ontologie,* Fourth Part, Berlin, 1934.

[4] Logicians will be quick to point out that this list is incomplete: one should at least add a (not a) and the empty set. These concepts have become even more problematic.

of ideal phenomena which used to be — and still largely are — equivocally reduced to their physical shadows. Unfortunately, among most of us this realization is still moored at a purely intellectual post. Only a few have carried out the readjustment of perception that would enable us to gaze at, say, moral or aesthetic values, inaccessible cardinals or quarks with the same objective detachment we adopt in gazing at the starry sky above us.

2. Once the thesis in (1) is admitted, one is led to two "correlative" views of the process of object-formation. First, only when an object is "taken seriously" and studied "at its own level" will it reveal its properties within its own eidetic domain. All scientists know this, more often irreflexively, and all scientific work is done in this spirit. But scientists are reluctant to adopt the obverse attitude: anything that has been objectified can just as well be "bracketed," and suddenly be seen as problematic.[5]

Every object carries within itself the seed of its own irrelevance. Thus for instance the development of mathematics has reached the degree of perfection where the notion of set is turning into a worn-out coin.

To get out of this impasse, Husserl developed *genetic phenomenology*. Briefly, anything that has been made into an object (in mathematics, we would say "defined") eo ipso begins to conceal the original drama that led to its constitution. Most likely, this drama followed a tortuous historical path, through things remembered and things forgotten, through cataclysms and reconstructions, pitfalls and lofty intuitions, before terminating with its objective offspring, which will thereupon naively believe itself to be alien to its origin.[6] The reconstruction of this genetic drama is a *logical process for which classical logic is totally inadequate*.

---

[5] Following Ortega's thesis that phenomenological reduction is a response to the problematization of a belief, expounded in *Obras Completas*, Madrid, 1955, V, 379-410, and esp. 544-547.

[6] This passage is partially adapted from E. Lévinas, "Reflexions sur la 'technique' phénoménologique," in *Husserl, Cahiers de Royaumont*, Philosophie No. III, Paris, 1959, pp. 95-107.

The examples that Husserl and other phenomenologists developed of this genetic reconstruction, admirable as they are, came before the standard of rigor later set by mathematical logic, and are therefore insufficient to meet the foundational needs of present-day science. In contemporary logic, to be is to be formal. It falls to us to develop the technical apparatus of genetic phenomenology (which, as Merleau-Ponty argued,[7] coincides with the much-looked-for *inductive logic*) on the same or greater a standard of rigor than mathematical logic. Again taking as an example the notion of set, one might begin by formalizing the ontologically primary relations of being which had to be veiled as soon as the two relations ⊆ and ∈ were constituted. The following are a few of them, largely taken from the phenomenological literature: *a* lacks *b*, *a* is absent from *b* (one could describe in precise terms how this differs from the classical "*a* ∉ *b*"), *a* reveals *b*, *a* haunts *b* (as in "the possibility of error haunts the truth"), *a* is implicitly present in *b*, "the horizon of *a*," and so on, and so on. Of great scientific interest is the relation of *Fundierung*, which ranks among Husserl's greatest logical discoveries.[8] The rigorous foundation of the concept of time provides further examples of such relations: as shown in Husserl's lectures on *Zeitbewusstsein*, the relations of object to past and future are irreducible to classical set theory, and lead to an entirely novel theory of impeccable rigor.

The stale objection that relations such as the ones listed above are "purely psychological" serves only to underscore the failure to adopt the attitude described under (1). If there is an evidence that phenomenology has conclusively hammered in, ever since the first volume of the *Logical Investigations*, it is that nothing

---

[7] In *Les Sciences de l'homme et la phénomenologie*, Paris, 1961.

[8] The only instance of such a formalization I know of is Alonzo Church's "A Formulation of the Logic of Sense and Denotation," in *Structure, Method and Meaning, Essays in Honor of Henry M. Sheffer*, New York, 1951, pp. 3-24. Unfortunately, Church's lead seems not to have been followed up, partly because the reading of his paper is a veritable obstacle course. We hazard the hypothesis that Husserl's Third Investigation could be subjected to similar formalization without excessive retouching.

whatsoever is "purely psychological." "*Tout est dehors*," as Sartre would say.[9]

3. The crisis of foundations is not limited to full-grown sciences such as physics and mathematics. As Husserl goes to great length to point out, it reappears under different guise in the more "retarded"[10] sciences, such as the life sciences and the social sciences, as well as the hypothetical eidetic sciences yet to be instituted (of which a typical list is found in *Ideas* I).

The relative lack of success of these sciences is obvious when measured by the most impartial yardstick, namely, the ability to make reliable predictions.[11] The culprit is to be found—it is Husserl who speaks—in the hasty adoption of the logic and methods of physical sciences, in the spirit we have briefly criticized under (1). Hypnotized by the success of physics, the newer sciences have failed to go through an independent Galilean process of concept-formation, which alone would have endowed them with an autonomous structure. The very real need to match the exemplary *rigor* of physical science was—and is— erroneously replaced by an uncritical imitation of physical methods and techniques.

Again, genetic phenomenology, this time with the purpose of construction (rather than dissection) of ideal objects to be subjected to yet-to-be-discovered ideal laws and relations, is the proposed remedy. Whether or not it will work depends on the inventiveness of scientists and philosophers now active in these areas.

From (2) and (3) we conclude that genetic phenomenology intervenes at two stages in the development of a science: at dawn, by circumscribing an autonomous eidetic domain with its internal laws; at dusk, by the criticism of that very autonomy,

---

[9] In the well-known note "Une Idée fondamentale de Husserl: l'intentionnalité," *Situations* I, Paris, 1947, pp. 31-35.

[10] This felicitous adjective is used in this sense by Jean Wahl in *L'ouvrage posthume de Husserl: La Krisis,* Paris, 1965.

[11] Lord Rutherford used to remark acidly that all science is divided into two parts: physics and stamp-collecting.

leading to an enlargement of the eidetic domain. The process can be schematically represented by the following recurring pattern:

$$\ldots \to \text{ideal object} \to \text{science} \to \text{crisis} \to \text{genetic analysis} \to \text{ideal object} \to \ldots$$
$$ \text{(a)} \phantom{\text{ideal}} \text{(b)} \text{(c)} \phantom{\to \text{cris}} \text{(d)} \phantom{\text{genetic analy}} \text{(a)}$$

It is left to us to develop this program in rigor and detail, with the help of the awesome amount of material Husserl left us. Should this program turn into reality, we may then live to see the birth of a new logic, the first radical reform of logic since Aristotle.

It is of course all too likely, unless communications between scientists and philosophers improve, that mathematicians and scientists themselves, unaware of Husserl's pioneering work, will independently rediscover the very same way out of their present impasse. Recall that an inspired genetic analysis of the concept of simultaneity, carried out by a philosophically untrained physicist, led to the creation of the special theory of relativity.

1975

CHAPTER FIFTEEN

Husserl

THE BEST PHILOSOPHERS OF OUR CENTURY suffer from a common deficiency of expression. They seem bent upon making an already difficult message all but unintelligible by irritating mannerisms of style. For example, in Wittgenstein we meet a barrage of epigrammatic cryptography suited only for the Oxbridge market; in Heidegger truth is subordinated to alliteration and to a cunning desire to anger the reader by histrionic displays of German archaisms; Ortega would bury his finest insights in prefaces to his friends' collections of Andalusian poems or in Sunday supplements of Argentine dailies, while feeding the grand public a dubious *Kitsch* calculated to keep himself financially afloat; Croce would use his pen to fly away from unpleasant Fascist reality into the anecdotes of the Kingdom of Naples of yore; Nicolai Hartmann was subject to attacks of graphomania; and so on, all the way to Sartre. Small wonder that the intellectual public, repelled by such antics, should fall into the arms of a demimonde of facile simplifiers and sweeping generalizers. The Russells, the Spenglers, the Toynbees, and their third-rate cohorts have lowered the understanding of philosophy to a level unseen since the seventh century.

The most pathetic failure in communication is Husserl. Acclaimed by many as the great philosopher of the century (Kurt Gödel ranks him as the greatest since Leibniz: Hermann Weyl held a similar opinion) as well as the most influential — most of the above considered themselves his students — he wrapped his thoughts in a heavy-handed, redundant, solipsistic German academic style which makes his writing all but impossible to translate and which has kept him all but unknown in the English-speaking world. David Carr[1] has accomplished an acrobatic feat in rendering the flavor of the German original of the *Krisis*, where doubts and hesitations alternate with bold assertions and eye-opening insights. The success of this translation, however, makes it all the more important to break the barrier to Husserl's thought. As consumers of philosophy we should like to have a brochure, a prospectus, before we decide whether or not to buy; hence this review.

I meet someone in the street. As he walks toward me, I recognize him: "It is Pierre." The Pierre I am meeting now is *the same* Pierre I met yesterday. How do I know he is *the same?* This question is usually given a mechanistic answer, by translating the event into physiological terms. My eyes register Pierre's image and transmit it to the brain, where in turn, by another physiological process, it is identified with yesterday's image.

But this identification cannot be performed without a further process taking place in the brain. Some neurons (say) must be endowed with the ability to recognize each other: The Pierre-yesterday-neurons must be able to recognize the Pierre-today-neurons. So, the original problem repeats itself unchanged at the neuron level, and a further mechanism is required to effect a second-order recognition. Proceeding in this way, we are led to what may be called a *regressio ad infinitum*. An infinity of mechanisms is required, the one at stage $n + 1$ designed to explain how at stage n recognition can take place.

By this argument (admittedly oversimplified) the mechanistic explanation of any phenomenon of perception is faced with an

---

[1] *The Crisis of European Science and Transcendental Phenomenology,* by Edmund Husserl, translated by David Carr, Northwestern University Press, Evanston, 1972.

impasse that cannot be evaded without intellectual dishonesty. The process whereby I recognize Pierre seems to defy physiological explanation. In a world where the only accepted explanation is a mechanistic one, such a conclusion is inadmissible, and in fact philosophers past and present have weaseled out of it by shrewd rationalizations. Descartes does not hesitate to invoke the intervention of a Higher Being, and Kant deftly avoids a direct confrontation. Scientists will be tempted to shrug it away as an irrelevant curiosity, unless forced to face it by such problems as pattern-recognition or machine translation.

Husserl's debut in philosophy is a head-on attack on this problem. Several hundred pages of the *Logical Investigations* and about half the text of the *Krisis* are dedicated to it and its variants, under the condemning epithet "psychologism" (a more common word nowadays is "reductionism"). To shorten the way to his main argument, we shall use an old distinction which enjoyed great vogue in the Middle Ages. A logical or causal interaction between objects or events in our experience may be classified as *material* or *formal*. Examples: The mind's physiology interacts with my mathematical work, but in a material rather than in a formal sense; my digestion, however important, is materially (*materialiter*) but not formally (*formaliter*) related to the theorem I am trying to prove. Beethoven's Ninth Symphony is formally the same, despite the variety of its material performances. We shall let the reader glean a definition from these examples.

The formation of a science results from singling out a set of concepts and laws which *formally* hold together, as well as a formal method of explanation, or *logic*, which renders such a science autonomous, that is, independent of all *material* relationships with any other domains of experience, despite the richness and usefulness of such relationships (as in so-called "applications"). The exemplary instances of this process are mathematics and physics, or mathematical physics for short. Husserl goes to great lengths, in the first part of the *Krisis*, to show how physics originated with Galileo by willful and artificial abstraction of a small set of concepts out of the chaos of medieval naturalism. The spectacular success of mathematical physics and its unmatched objectivity result from the understanding — now instinctive among scientists — of the formal method which allows the discipline to progress on its own steam, coupled with the evidence — now taken for granted — that no material interaction with any other happenings can formally affect such progress, though it can do so materially, for example by the withholding of research funds.

For Husserl the failure to produce a formal explanation, or, what is more serious, the substitution of a material explanation where a formal one is called for, is a major source of error. If logic is to be primarily the theory of error, as Hegel wrote, then this assertion can be taken as the key to Husserl's logic.

We can now return to Pierre. The physiological explanation is a material, not a formal one. Unless a formal explanation is given, we shall never get very far in patching up the *regressio ad infinitum*. We must view Pierre-across-the street independently of psychological considerations, as an event-in-the-world to be investigated, compared with related events, and eventually fitted into a formal theory. In my perception of Pierre, his sameness-as-yesterday may materially be a happening of my neurons, but formally it is the manner by which Pierre appears, structurally similar to, say, his appearing-across-the-street, barely-glanced-at, or surprisingly-early. These manners of appearance are properties of Pierre, deserving as much objectivity as his weight or the color of this hair. Our bias in granting the gift of objectivity stems from an age-old and no longer tenable prejudice that singles out physical objects as the only real ones and dismisses all other domains of experience by material reductions. I shall have to start studying, with the same thoroughness as a chemist does with his elements, the properties of that particular form of being which is "sameness-as-yesterday," its logic and its laws, in short, its *theory*. If successful, such a theory will make the physiological explanation of my recognizing Pierre appear as preposterous as a physiological explanation of the existence and uniqueness of the five Platonic solids. Outrageous as it may sound, this kind of reasoning is precisely what Husserl success-fully pulls off in several instances, most strikingly in his astonish-ing analysis of time.

By the example of Pierre we are led to Husserl's main con-tention. "All being is formal being." Every object of experience must have a formal *correlate* and must enjoy at least the possibility of being ensconced into a formal domain. The word "formal" is here very close in meaning to "objective," except that the latter is tainted with the presumption that the object under study is physical. To Husserl some of the most intriguing sciences-to-be are to be found in areas which are now dismissed as "purely psychological." Nothing whatsoever is purely psychological, he affirms. Everything, even remembrance or a mental image, must, by inescapable logical necessity, have an objective cor-relate which can be formally studied. The persistent prejudice that the weight and color of an object are worthier of scientific investigation than its absence or its purpose is to Husserl a dictatorial absurdity, inherited from a materialistic past, whose destructive effects upon science and society are becoming all too apparent.

To be sure, "formal" is not here to be meant to identify with mathematics or formal logic as they are conceived today. To avoid this misunderstanding Husserl chooses another slogan for his main contention: "Consciousness is always consciousness of something." The meaning is the same, but harder to decipher, unless one is privy to his verbal games. Unfortunately, this iden-tification is frequent in psychological and social studies, and it reappears more subtly in biology and medicine. In their search for a formal structure, the non-physical sciences rush into a hasty adoption of mathematical or physical methods, and we find nowadays a horrors' gallery of "social physics" and "mathematical biologies" which gain temporary acceptance and only delay the construction of an autonomous foundation. The Galileo of biol-ogy is yet to come, and without him the methods of biology will not appreciably differ from those of a stamp collector.

It is legitimate to ask what Husserl means by formalization, if it isn't to be related to logic as conceived today. We meet here one of Husserl's most daring proposals: the reform of logic. The foundations of our logic were set by Aristotle and have remained unchanged since. That magnificent clockwork mechanism that is mathematical logic is slowly grinding out the internal weaknesses of the system, but it will never revise its own foundations without an impulse from outside. Present-day logic is based upon the notion of *set* $(A, B, C, ...)$ and upon two relations between sets: "A is an element of B" and "A is a subset of B."

We must soberly remind ourselves that these notions are not eternal and immutable, but were invented one day for the purpose of dealing with a certain model of the world. What if this model were inadequate to the needs of the new sciences? To expect that a logic conceived for the description of physical and inanimate objects should be adaptable to the infinitely subtler logical requirements of theories of living organisms — to name but one instance — is nothing short of blindness. If we are to set the new sciences on firm, autonomous, formal foundations, then a drastic overhaul of Aristotelian logic is in order. This task is far more complex than the Galilean revolution. Pre-Galilean physics was by and large a failure. But mathematical logic is a wildly successful enterprise in its avowed aims, and it is ingrained as second nature in our minds.

Husserl begins his attack by digging out of ordinary experience a number of crucial relations-of-being which mathematical logic ignores. He then attempts or at least proposes their formalization. Examples: "A is absent from B," "A is *already* contained in B," "A anticipates B," "A is a perspective (*Abschattung*) of B." Past, present, and future, conceived as formal, no longer psychological, relationships, become intriguing logical problems calling for a totally novel concept of *ens*.

It is tempting to be drawn to enthusiasm by the cogency of Husserl's arguments and to conclude that other formal sciences will soon be born, in parallel with mathematical physics. Unfortunately, we know better of the ways of the world, and so did Husserl. While writing in seclusion amid the horrors of the thirties, he tried to convince himself, with this book, that Western culture could be saved from the disaster that befalls a man who has overdeveloped one limb at the expense of others. In opposition to the naive materialism of our fathers, he proposes a countermeasure which is free from the booby-trap of irrationalism. As Sartre has remarked, Husserl's philosophy "gives us a great feeling of liberation. Everything is outside; we need no longer consider phenomena like knowledge and perception as being akin to the swallowing of food."

Whether and to what extent Husserl's revolutionary proposals will be implemented will depend largely on the work that is being done now, on the superior quality of translations (such as this one), and on the circulation, re-elaboration, and acceptance of his ideas in scientific circles. We have presented only one thread out of a mass of original ideas, suggestions, and brilliant visions which, like a large quarry, may provide the marble for many a future theory. Let us hope he will be spared the fate, reserved to other philosophers, of being mummified by scholars and enshrined in a marble mausoleum.

1974

CHAPTER SIXTEEN

Artificial Intelligence

Origins

SCIENTISTS DOING RESEARCH ON ARTIFICIAL INTELLIGENCE are still far from their ultimate goal — a computer-based analog of the human brain. Nevertheless, their efforts have regularly produced results useful in applied computer science. For example, the design of programming languages in general use has benefited from concepts developed in work on artificial intelligence.

Many branches of science have profited from the reductionist approach — the dissection of complex phenomena into elements whose simple interactions account for these phenomena. This approach is particularly characteristic of computer science, which uses dissection as a universal technique for system design. For example, to produce a full-blown graphic display system, a software designer will first analyze the geometric functions to be provided into vector and matrix operations which can represent these functions. Then he will decompose these vector and matrix operations into standard arithmetical, data-movement, and condition-testing operations and, if necessary, decompose these simple standard operations into absolutely elementary patterned openings and closings of transistorized switches. This method of systematic reduction can be applied now with considerable sophistication; sometimes as many as a dozen intermediate conceptual layers appear between the elaborate functions that certain computer systems provide and the billions of elementary switching steps on which these functions ultimately rest.

From the start, this habit of dissection, coupled with the realization that the computer can manipulate perfectly arbitrary information patterns, has suggested that the broad complex of abilities that constitutes human intelligence might itself be dissected and then reconstructed artificially. To do so would be to build an artificial intelligence, a system of programs that, when run on a sufficiently powerful computer, could imitate all of the intellectual capacities of the human brain: appropriate responses to spoken language, visual perception and manipulation of objects in three dimensions, and even the ability to plan successfully in an environment of varied and complex contingencies, to invent new mathematics and new science generally, to play all games brilliantly, and to converse with wit and verve in many languages.

Potential Impacts   The construction of artificial intelligence would affect the circumstances of human life profoundly. The appearance of intelligent beings other than man would surely create a new economics, a new sociology, and a new history. Moreover, if artificial intelligences can be created at all, there is little reason to believe that they could not lead swiftly to the construction of artificial superintelligences able to explore significant mathematical, scientific, or engineering alternatives at a rate far exceeding individual human ability. Such expectations motivate many computer scientists, whether they concern themselves directly with computer realization of some intelligent function or with efforts such as computer systems design, development of programming techniques, or invention of specialized algorithms. The possibility has been mentioned often in wider circles, but public discussion has not begun to reflect this perspective in any adequate way.

The efforts of a growing body of determined researchers in artificial intelligence have produced real successes, and the general scientific and industrial influences of their work are steadily increasing. Nevertheless, the basic techniques available to the worker in artificial intelligence are still modest.

Current Status

To assess the current situation, we can contrast the techniques of programming with the more ambitious goal of learning. A computer is programmed by supplying it with a carefully composed sequence of instructions which guides its actions in all necessary detail. But to learn, a computer would have to ingest relatively unstructured masses of information, much closer to the highly fragmented information that people deal with routinely, and itself supply the elusive steps of error correction and integration needed to turn this broken material into polished instructional sequences which can be followed literally. Only to the extent that a computer can absorb fragmented material and organize it into useful patterns can we properly speak of artificial intelligence.

Researchers in artificial intelligence have therefore sought general principles which, supplied as part of a computer's initial endowment, would permit a substantial degree of self-organization thereafter. Various principles that allow useful structures to be distilled from masses of disjointed information have been considered candidates in this regard. These include graph search and deduction from axioms.

Graph Search Many problems can be reformulated as one of finding a path between two known points within a graph. Planning and manipulation problems, both physical and symbolic, illustrate the point. Such problems are described by defining (1) an initial condition with which manipulation is to begin, (2) some target state or states that one aims to reach, and (3) a family of transformations that determines how one can step from state to state.

The problem of chemical synthesis is an example: the target is a compound to be synthesized, the initial state is that in which easily available starting substances are at hand, and the allowed manipulations are the elementary reactions known to the chemist. The problem of symbolic integration is a second example:

a given formula F containing an integral sign defines our starting state, any formula of the class mathematically equivalent to F but not containing an integral sign is an acceptable target, and the transformations are those that mathematics allows.

In all such problems, the collection of available transformations is a heap of relatively independent items, since transformations can be listed in any order, any arbitrary collection of transformations defines a graph, and a collection of transformations can be freely expanded or contracted without the appearance of any formal obstacle. Thus, the construction of a path through the graph defined by a collection of transformations does represent a situation in which a structured entity, a path, arises via a general principle from something unstructured and indefinitely extensible, a collection of transformations. One can try, therefore, to use this construction as a universal principle for the automatic derivation of structure.

Deduction from Axioms The work of numerous mathematicians in the nineteenth and twentieth centuries has shown that the classical corpus of mathematics can be based on a handful of astonishingly simple axioms and inference rules. The resulting formalism, the logician's "predicate calculus," is a straightforward language of formulas, in which any assertion of classical mathematics can be stated readily. In this system, a simple and easily programmed set of rules suffices to define all possible mathematically justifiable lines of reasoning. Thus, the predicate calculus gives us a simple but universal formal framework encompassing all logical reasoning.

Arbitrary collections of axioms that express the laws of particular mathematical and nonmathematical domains can be written readily in this calculus; then, if we are willing to search a sufficiently large space of proofs, the consequences of these axioms can be extracted automatically. Since axioms can be listed in relatively unstructured fashion, we have here another situation in which highly structured entities — mathematical proofs — arise via a general principle from something unstructured and indefinitely extensible — a collection of axioms. This construction clearly can be used as a universal principle; a considerable amount of work has been done toward adapting this fundamental logical mechanism to other uses, such as the automatic generation of plans and programs. Generally speaking, such adaptation proceeds without difficulty.

The Efficiency Problem

Nevertheless, neither the use of this very general logical machinery nor of the graph-searching technique can be regarded as more than a fragmentary key to the problem of automatically developing significant structures out of fragmented masses of information. The reason is the overwhelming inefficiency of these methods. Unless care is taken, any attempt to search the space of all proofs possible in the predicate calculus will founder almost immediately because of the immensity of this space. Similarly, representation of significant problems such as graph searches generally requires consideration of a graph far too large for the largest imaginable computer.

This realization has inspired two decades of effort to prune the searches needed to find either significant proofs in logic or paths or other interesting combinatorial substructures in graphs — that is, to find ways of distinguishing profitable from unprofitable directions of exploration in order to increase the efficiency of these searches by the very large factors necessary. Several general and quite useful pruning principles have been found, but even the very best general proof-finding and graph-searching techniques known currently are incapable of handling more than small examples.

From this disappointing fact, many researchers have drawn the conclusion that significant progress will be possible only if the sought-after, structure-generating mechanisms somehow use not only very general techniques, such as those reviewed above, but also use larger amounts of information specific to particular domains. However, it is not at all clear how such information can be represented or organized best. Moreover, unless one is careful, the use of such information can be considered a covert retreat from the basic goal: general principles allowing self-organization. Indeed, the presence of enough manually supplied, preorganized information can lead us to classify a system simply as a clever program rather than as a recognizable step toward constructing artificial intelligences.

Rule Systems   This objection need not apply to systems to which information can be supplied in the form of separate rules of the kind that an expert in some area might use to transmit his expertise to a beginner. For this reason, such rule systems have begun to attract attention recently. Attempts have been made to use large collections of rules to build artificial expert systems, such as programs able to diagnose human or plant diseases or to recognize clues to the presence of minerals.

Once rules have been supplied to such systems, the systems ingest lists of specific clues or symptoms visible in the situation to be assessed. To each clue or symptom, a numerical indication of the degree of reliability may be attached. The rules specify how these clues are to be combined to yield estimates of the likelihood that particular expert judgments — diagnosis of a specific disease or detection of a mineral — are correct. If the estimated likelihood is high enough, the corresponding judgment is pronounced.

In simple rule systems of this kind, computation of the relevant internal estimate may involve little more than a polynomial combination of the presence/absence/reliability indicators for the clues. When this is so, the rules merely define the coefficients of the polynomials or other functions that will be used. Even in such simple cases, the system may provide a specialized language for defining rules that allows the experts — for example, physicians or mineralogists — from whom the rules are to be acquired to express themselves in a language more familiar to them than the mathematical language of polynomials and functions.

The estimation procedures used in the simpler expert systems of this kind are not very different from some of the mathematical grouping techniques used with limited success for many years to catalog and look up journal articles and other scientific citations automatically. These procedures are admittedly crude, but evidence is beginning to appear that they may not be too unrefined for practical application. (Conceivably, their crudity may reflect that of some of the internal processes that a human expert uses to form a snap judgment of the evidential value of combinations of familiar clues.)

Links with Applied Computer Sciences

In one sense, the rule systems, the graph and transformation systems, and even the logical formalism that workers in artificial intelligence have tried to use are simply specialized programming languages. The particularly interesting point about these languages is that they aim, in a more single-minded way than ordinary programming languages, to decompose the programs they express into small, independent fragments. As emphasized above, the ability to deal with fragmented material and integrate it automatically is a fundamental goal of research in artificial intelligence. Languages that make it possible to use small independent fragments to define complex processes are of great interest to programmers even outside the field of artificial intelligence. Such languages eliminate a most troublesome source of programming error and can increase programming speed very considerably. For this reason, research in artificial intelligence has been a fertile source of concepts that have passed into the design of generally used programming languages.

Another connection between work in artificial intelligence and the more technical part of computer science can be noted. To describe and analyze the syntactic structure of programming languages, computer scientists customarily use a kind of formal mechanism called a context-free grammar. Grammars of this kind specify how statements in a programming language are to be parsed (a preparatory operation very close to the sentence diagramming taught in high school English classes a generation ago).

These formal grammars were invented more or less simultaneously by workers in computer science and students of natural language. Although simple, they are remarkably effective in capturing, at least roughly, the general grammatical constructs that people seem to comprehend easily and naturally. Variants of this same formal mechanism have been used in systems for analysis (and, to a limited extent, understanding) of natural language and in systems for decoding spoken utterances. Although context-free grammars are used in many areas and are not always thought of as belonging specifically to artificial intelligence, they do represent a genuine case in which significant aspects of an important human function can be captured by a simple computer model.

Expectations

Research in artificial intelligence is still immature. As yet, it can boast of few significant general principles and of little theory capable of directing the work of its practitioners. Nevertheless, the field is not without its successes. In particular, the expectation that artificial synthesis of intelligence will be possible has inspired attempts to computerize functions, such as the decoding of spoken utterances and analysis of visual scenes, of a subtlety that might have seemed unreachable otherwise. Full duplication of human sensory and intellectual capabilities, which are rich and subtle indeed, still lies far beyond our reach. But attempts to match them are beginning to be successful enough to furnish items valuable as components of practical systems.

It should be noted that the efforts at artificial vision, while still far short of the full capabilities of human vision, can have a major impact when joined with particular developing industries. An example is the work on robotlike automatic devices — now generally being called industrial robotics — which has much to contribute to industrial productivity. The artificial intelligence work in vision is beginning to add substantially to the capabilities of these devices and will play an important role in the development of the field during the coming decade.

Seen from a more fundamental point of view, the most successful applied programs that have grown out of the work on artificial intelligence may be judged to exhibit only limited degrees of self-organization but nevertheless to represent the application of ingenious programming to unusual reas. Such fragmentary successes continue to accumulate. We may expect that they will cohere gradually suggesting eventually more adequate general principles of self-organization than are now available.

1982

CHAPTER SEVENTEEN

Computing and Its History

ALL REVOLUTIONS, like the computer revolution, are described and recorded before they happen. While computers were still inconspicuous adding machines, star-gazers and senior scientists took turns at providing unsolicited accounts of the New Age. But as soon as the revolution began to affect our lives, predictions on the shape of science and society in the age of the computer became cautious and rare.

The art of prediction, whether practiced by professionals or by amateurs, deals with the discontinuity of change. In the face of a staggering jump never before encountered, the lessons of yesterday are only logarithmically significant, and the achievements of the past are no longer the source of inspired guesswork. Today's accomplishments in computing beat those of our immediate predecessors by several orders of magnitude.

Yet, now more than ever the time calls for a cool assessment of the effects of the computer revolution. One need not wait for the next major change, for the advent of massive parallel systems now at the planning stage, in a world race so feverish as to call into question the work ethics of East and West. Even the media are now broadcasting the opinion that the future of civilization no longer depends as much on armies and weapons as on arrays of supercomputers, and on the inspired software that the best minds of each nation will devise for them. The stakes are high. As in old science fiction novels, the reward will go to whoever is first to program in the four dimensions of space and time.

Every corner of science will then be revamped. The predictive powers of quantum mechanics — now limited to small atoms — and those of mathematical economics — now limited to past events — will be triumphantly verified or definitively refuted. Statistical mechanics will explain why water boils at 100°, and biologists will be experts in combinatorics. Artificial intelligence will be pushed to its natural limits, and the uneasy boundary between mind and matter will be drawn. The mathematician, far from being thrown into the dustbin of history, as some mistakenly fear and others secretly expect, will again be called upon to explain extraordinary phenomena to be revealed by experimentation with the computer.

In fact, the demands on the powers of the intellect, far from being taken over by computers, as some simplistically predict, will be greater than ever. All trite and routine work removed, the scientist will be forced to face those tasks that call for the exercise of his creative faculties, those he does not share with machines. Even scholars in the humanities, freed from the trivia of erudition, will return to their calling as men of letters, and the new music may well surpass that of the Baroque age, as composers experiment on their terminals with heretofore unheard permutations of sound.

As economic planning is made effective by computing, the world will benefit from the demise of political ideologies based on wishful thinking. And not a minute too soon, as the depletion of natural resources will threaten a famine that only computer-aided foresight will avert. Similarly, only a swift pharmacology that will bypass lengthy experimentation shall prevent mankind from falling victim to the new diseases that are already darkening the horizon.

The coming generations will at first rejoice, as every child acquires analytic skills that are now a privilege of the technological elite. Instruction by computer will weed out incompetent teachers and old educational shibboleths. But minds that will be largely trained in analysis will be made aware of their deficiencies in synthetic and speculative thought. Will a new kind of teacher come along who will meet these unexpected educational needs?

We shall live in a society of poverty and plenty. Communication on the computer's display will make ordinary conversation rare and difficult, but no less coveted, and the flood of impersonal information will stress the lack of genuine human contacts. As everyone becomes skillful at dealing with abstract entities, society will suffer from clumsier human relations. The survivors of our age will then be sought-after wise old men, and our time may then be admiringly studied for clues to a lost happiness. *Historia* will then again, with a vengeance, be regarded as *magistra vitae*.

1985

CHAPTER EIGHTEEN

Will Computers Replace Humans?

I should say that most of the harm computers can potentially entrain is much more a function of properties people attribute to computers than of what a computer can or cannot actually be made to do.

Joseph Weizenbaum, in *Science,* May 12, 1972

THE COMPUTER, one can safely predict, will be adjudged to have been the ultimate technological symbol of our century. Although not nearly as common as a car or a TV set (the runners-up in the race for the ultimate technological symbol), it has affected our views, our attitudes, and our outlook in more subtle and disquieting ways.

We find it difficult to accommodate ourselves to the computer because, on the one hand, it has proved to be an indispensable tool in such great adventures as that of putting man on the moon and, on the other hand, it threatens our fundamental rights by its ability to keep and to hold ready for instant recall prodigious amounts of information about our activities, our habits, and even our beliefs.

Every gasp of admiration for the remarkable feats it can perform is matched by a groan of pain caused by its apparent inability to shut off a stream of dunning letters long after the bill has been paid.

Never has there been an instrument of such capacity for the useful and good and at the same time of such potential for mischief and even evil. And what makes our coming to terms with the computer so difficult and so frustrating is that we find ourselves face to face with a being seemingly capable of some imitation of the highest of human acts, namely that of thought, while at the same time totally devoid of all human qualities.

"To err is unlikely, to forgive is unnecessary" is the computer's version of the old proverb that "to err is human, to forgive divine." This is frightening especially if one thinks of the antics of HAL, the deranged computer of "2001."

We might be less frightened if we knew a little more about modern computers: what they are, what they can do and, perhaps most importantly, what they cannot do.

Computers of one kind or another have been with us for a long time. The abacus came from antiquity and is still in use. The slide rule is probably close to being two hundred years old and today is no less indispensable to scientists and engineers than it ever was. The adding machine has been in use for well over half a century and such "business machines" as sorters and collators still account for a significant proportion of IBM rentals.

Throughout history, man has repeatedly tried to relegate the drudgery of computing to a machine and to increase the speed and accuracy of routine calculations. Perhaps the most remarkable of all such efforts was the work by Charles Babbage (1792-1881), because in many ways it foreshadowed the advent of modern electronic computers. Babbage's inspired and clever "calculating engines" came a bit too soon, for neither the minds of his contemporaries nor the technology of his day were quite ready for them.

The present-day, all purpose digital electronic computer (to give it its rightful name) is a product of the unlikely union of advanced electronic technology with some quite abstract developments in mathematical logic.

The spiritual father of the modern computer is the late A. M. Turing, a highly gifted and original English mathematician, who in 1936 conceived of a "universal machine" that now bears his name.

The machine consists of an infinite tape divided into identical squares. Each square can either be blank [which is denoted symbolically by (*)] or have a vertical bar (|) in it. There is also a movable scanning square that can be moved either one step to the right or one step to the left.

The only operations allowed are:

L : move the scanning square one step to the left
R : move the scanning square one step to the right
* : erase the bar (|) (if it is there)
 | : print the bar (|) (if it isn't there)

Instructions to the machine are in the form

$$5 : * \mid 7 ,$$

which is read as follows:

Instruction 5 : If scanning square is over a blank square, print |
and see instruction 7,
Instruction 7 may read as follows:

$$7 : \mid R\ 7 ,$$

i.e., if the scanning square has a vertical bar in it, move it to the right and repeat instruction 7. However, if, after following instruction 7, the scanning square is over a blank square, then either there is a second part to instruction 7, e.g.,

$$7 : * L\ 8 ,$$

or there is no such second part, in which case the computation terminates because the machine cannot go on. The Turing machine owes its fundamental importance to the remarkable theorem that all *concrete* mathematical calculations can be programmed on it. ("Concrete" is to be understood in a precise technical sense, but this would involve a discussion of so-called recursive functions and take us a bit too far afield. I here choose to rely on the reader's intuitive feeling for what is meant by concrete.) In other words, every concretely stated computational task will be performed by the machine when it is provided with an appropriate, finite set of instructions.

Because the machine is so primitive, it requires considerable ingenuity to program even the simplest task. For example, multiplication by 2 is accomplished by the following set of instructions:*

$$
\begin{array}{l}
0 : \mid\ *\ 1 \\
1 : *\ R\ 2 \\
2 : \mid\ *\ 3 \\
3 : *\ R\ 4 \\
4 : \mid\ R\ 4 \\
4 : *\ R\ 5 \\
5 : \mid\ R\ 5 \\
5 : *\ \mid\ 6 \\
6 : \mid\ R\ 6 \\
6 : *\ \mid\ 7 \\
7 : \mid\ L\ 7 \\
7 : *\ L\ 8 \\
8 : \mid\ L\ 1 \\
1 : \mid\ L\ 1
\end{array}
$$

It should be understood that a number (n) to be multiplied (by 2) is represented by $n + 1$ consecutive vertical bars, and the computation is begun by placing the scanning square over the bar farthest to the left (instruction 0 starts the calculation).

Figure 2 shows eight consecutive stages of the calculation for $n = 3$, and the reader should find it instructive to continue to the end to discover the answer to be $3 \times 2 = 6$.

A real computer is, of course, much more complex and also a much more flexible instrument than the Turing machine. Nevertheless, the art and science of programming a real computer, although sometimes impressively intricate, is very much like that of instructing a Turing machine.

As the reader will have noticed, the instructions must be extraordinary explicit and complete. The slightest error ("bug") and

---

* Taken from *Theory of Recursive Functions and Effective Computability* by Hartley Rogers, Jr., McGraw-Hill Book Co. 1967.

the result will be hopelessly wrong, the machine will fail to turn itself off, or some other disaster will happen. In communicating with the computer there is nothing like the "don't you see?" or "think a little harder" sort of thing. And there is no point whatsoever in getting angry or upset. It will only cause more errors and the computer will go on printing out more gibberish which it "thinks" it was instructed to print out.

Programming calls for a most detailed and rigorous analysis of the task at hand, and "debugging" is puzzle-solving on the highest level.

The difficulties and subleties of programming are caused to a large extent by the primitiveness of the basic operations that the computer can perform. Could one not improve matters by enlarging the set of fundamental operations? Unfortunately, what might be gained in simplicity and flexibility would be lost in accuracy.

Operations like L, R, or "erase" and "print" are "off-on" operations that require only a flip of a mechanical relay or the opening or closing of an electrical circuit. The only limitations for such operations are cost and speed, and in principle they are infinitely accurate. (This is also the reason for preferring base 2 in computer calculations. To this base, the digits are 0 and 1, ideally suited for "off-on" devices.) Mechanical relays are cheap but also are much too slow, and it was therefore not until they could be replaced by their electronic counterparts that the age of the computer really began.

Actually, the main stumbling block was "memory," i.e., a method of storing information and instructions in a way that would allow for rapid access and recall. This problem was first overcome in the mid-forties with the invention by F. C. Williams of a special memory tube. Improvements and new inventions followed in rapid succession, continuing until this day.

A modern computer is thus a device that can perform extremely simple operations at a fantastic speed and that is endowed with a "memory" in which vast amounts of readily (and rapidly!) available data can be stored for immediate or future use. Instructions are also stored in the memory, but before the

199

computer accepts them ("understands them") they have to be written in one of several special "languages," each designed for a particular use. Some of the most familiar of those languages are FORTRAN, A.P.L., and B.A.S.I.C. This, in essence, is all there is to it. Of course, the detailed way in which everything is put together is intricate and enormously complex.

It would appear that the computer is completely dependent on the man who instructs it. "Garbage in, garbage out" is the contemptuous way in which the computer's total lack of initiative or inventiveness is sometimes described. But this view is not quite just, and the computer can "teach" its masters things they may not have known. To illustrate this I shall recount a story which, although to some extent aprocryphal, is entirely plausible.

Some time ago, attempts were made to instruct the computer to prove theorems in plane geometry. To do this, in addition to giving the computer the axioms and rules of logical inference, one clearly had to provide it with a "strategy." The strategy was simple and consisted of the sole command: "Look for congruent triangles."

Given a theorem to be proved, the computer would first consider all pairs of triangles and, by going back to the assumptions, try to ascertain which of these (if any) were congruent. If a pair of congruent triangles was found, then all conclusions from this discovery would be drawn and compared with what had to be demonstrated. If a match resulted, the computer would proudly print "Q.E.D." and shut itself off. If no conclusion drawn in this way matched the desired one, or if no two triangles were congruent, then the computer was instructed to drop all possible perpendiculars, draw all possible bisectors and medians, and start to search for congruent triangles all over again.

Well, here is what happened when the computer was asked to prove that, given an isoceles triangle with $AC = BC$, angles A and B are also equal [$\angle A = \angle B$].

The authors of the program fully expected the computer, after a brief contemplation of the lonely triangle ABC, ultimately to drop the perpendicular CD^* and, from the congruence of the right triangles ADC and BDC, conclude that $\measuredangle A = \measuredangle B$.

But the computer fooled them! It gave a much nicer and a much more sophisticated proof. It simply "noted" that triangles ABC and BAC (!) are congruent and derived the desired conclusion in this way.

Before anyone jumps, let me hasten to state that this proof is not new. In fact, it is very old and was first given by Pappas in the fifth century A.D. It was reproduced in most texts on geometry until about the end of the nineteenth century, when it was decided that it was too difficult for young humans to comprehend.

The reason the proof is difficult is that it goes against the grain to consider ABC and BAC are *different* triangles, even though they are represented by the same drawing.

Intuition based on sensory experience is sometimes a deterrent to abstract thinking. Therefore, because a computer is deaf and blind, so to speak, it can in some respects and in some instances win over humans.

However, can a computer be truly inventive and creative? This is not an easy question to answer, because we ourselves do not really know what inventiveness and creativity are. One of the more interesting and imaginative lines of research, which goes by the name of "artificial intelligence," is directed toward answering such questions and, somewhat more generally, toward computer simulation of activities usually associated with living beings which may even be endowed with intelligence. How complex the notion of creativity really is can be illustrated by the following example:

Consider an eight-by-eight-inch board made out of equal (one-by-one inch) squares from which the left uppermost and the right lowermost spaces are removed.

---

*Students of "New Math," rejoice! The proof requires that D is between A and B, which order axioms are needed.

Given a set of two-by-one "dominos," the question is whether one can completely fill the board by placing them horizontally and vertically in any position without overlapping.

The answer is no, and the proof is as follows:

Imagine that the squares of the board are colored alternately like a chess-board (e.g., white and black) so that any two adjacent squares differ in color. Of the 62 squares, 30 will be of one color and 32 of the other. If we also imagine that one square of the "domino" is black and the other one white, then the set of squares that can be covered by the dominos must, by necessity, have as many squares of one color as of the other. Since our truncated board contains unequal numbers of differently colored squares, it cannot be completely covered by "dominos," q.e.d.

The creativity comes in thinking of coloring the squares with two colors (or, more abstractly, of picking two symbols and assigning differing symbols to alternative squares). There was nothing in the problem as posed to suggest coloring (except perhaps that an eight-by-eight board suggests a chess-board), and the device of alternate colors had to be *invented* for the purpose of proof. Could a machine hit upon such an idea? Even if, by accident, it were to hit upon assigning symbols to squares, would it recognize the performance of this device to the problem as posed?

The enthusiasts of "artificial intelligence" would unhesitatingly answer in the affirmative. They believe that even a spark of genius is a matter of accumulated experience, and that therefore a machine that has a large enough memory and is capable of extremely rapid collation could, on occasions, pass for a genius. Well, perhaps. But that would not be of much interest if it turned out that, to be creative on the human level, the machine would have to be of the size of our galaxy — or even larger.

As of now, the computer has not even been really successful in performing intellectual tasks that should not be overly taxing. For example, machine translation of relatively simple texts from one language into another have not gone beyond a rather primitive stage, in spite of much effort in this direction. Theorem-proving, despite the (alleged) rediscovery of Pappas's proof, has not progressed far, either, and it may be some time before a computer can compete with a really bright young student.

Still, attempts to simulate feats of human intelligence by machines are interesting and worthwhile, if for no other reason than that they should help us to understand better the workings of our own minds.

However, the list of things the computer can do, and do well, is enormous and it grows longer by the day. Virtually every area of research depends to some degree on a computer. One can now begin to solve realistically approximate equations that describe the motion of the atmosphere. One can simulate in the computer the workings of telephone exchanges, and we should be able to avoid future blackouts affecting large areas by computer studies of the flow of electrical engery in complex networks. Calculation of orbits is routine, and so are many other calculations in physics and astronomy. Space studies are possible because of the vast amount of information poured into computers by satellites, unmanned probes of deep space, and manned space vehicles. Geologists use computers to analyze seismographic tracings that could be forerunners of earthquakes.

In the life sciences, the computer is now an indispensable tool for analyzing X-ray diffraction patterns, which are fundamental in determining structures of such complex molecules as DNA or hemoglobin. It processes huge amounts of neurophysiological data and it helps in testing all sorts of hypotheses and theories in ecology and population genetics.

There are manifold uses for the computer in economics, and in recent years attempts have been made to use it on a grandiose scale to predict the economic future of the world. Although one may question the assumptions underlying the proposed model and remain sceptical about the validity of conclusions drawn from it, the fact remains that an exercise of such magnitude could not have been even contemplated without the computer.

Here, however, a word of caution is necessary. It is precisely because of the power of the computer to simulate models of great complexity and because, in the popular mind, the computer is the ultimate in precision and infallability, that special care must be exercised in warning that the conclusions drawn are merely consequences of the assumptions put in. It is perhaps here that the criticism "garbage in, garbage out" may be justified, unless the assumptions are subjected to a most rigorous and critical analysis.

I could go on and on reciting the uses of computers in engineering, business, and government. There is, in fact, hardly any facet of our lives which has remained untouched by the rapid development of this instrument. But, instead, I should like to conclude by a few remarks about the "intellectual" role of the computer.

It is a difficult, and even a dangerous, task to try to delineate achievements from the instruments that made these achievements possible. Biology and astronomy would not be what they are today if it were not for the microscope and the telescope. And yet neither of these magnificent instruments is what might be called, in an "intellectual" sense, a part of the discipline it serves so indispensably and so well.

It is different with a computer. Although I cannot think of a single major discovery in mathematics or science that can be attributed directly to the computer, in an intellectual sense, the computer as such belongs, without doubt, to mathematics. This is not only because it is rooted in logic, but mainly because it has become a source of new mathematical problems and an instrument of mathematical experimentation. One could even attempt to defend a thesis (although it would be considered ultrasubversive by many, if not most, of my colleagues) that the only mathematical problems which arose during our century and are not traceable to the nineteenth are conceptual problems generated by computers.

The enormously complicated calculations which had to be performed during the wartime development of the atom bomb led John von Neumann, one of the great mathematicians of our time, to design and build one of the first all-purpose electronic computers. His vision, perseverance, and persuasiveness are primarily responsible for the flowering of the computer industry today. But von Neumann was also aware of, and fascinated by, the new world of automata and the novel problems they pose. He predicted, not entirely in jest, that the computer would keep scientists "honest," because logic is a prime requirement in the formulation of a problem that would have meaning for a computer. That is, unless the scientist is himself logical, the computer will be of no help to him at all. And in an intellectual tour de force, he invented a way in which an automaton could reproduce itself (i.e., be capable of reconstructing its own blueprint). Prophetically, this feat bore a more than perfunctory resemblance to the discovery some years later of the way in which living cells accomplish this task.

If computers can inspire such flights of imagination, they can and should be forgiven for some of the silly, irritating, and frustrating things they sometimes do when guided by their imperfect human masters.

1974

CHAPTER NINETEEN

Computer-Aided Instruction

Introduction

TODAY'S RAPIDLY ADVANCING ELECTRONIC TECHNOLOGY is working profound changes in the economics of information storage and information display. The hardware necessary to create attractive, inexpensive, widely disseminated, computer-based educational environments is coming into being.

While video games have grown into a five billion dollar industry, computer-aided instruction, which is supported by many of the same technological trends (including cheap microprocessors and high-quality graphics), is still a largely static area. Two main reasons for this difference can be asserted:

(A) CAI systems developers have not made effective use of the exciting, fast-paced graphics environment central to the video game industry. Partly because the equipment available to CAI developers has been inferior, typical CAI systems do not integrate all the facilities that would allow for a compelling multimedia educational environment (color, voice and music output, video images, split-screen and multi-screen displays, etc.)

(B) The cost of creating high quality courseware is still very high, and because of this, most currently available courseware is mediocre. The cure for this problem will require a substantial up-front investment in improved courseware-writing tools. (Some of what is wanted is outlined in Section 4.) One must also expect production of really high quality software to require TV-studio-like teams incorporating many specialists: subject-area experts, talented teachers, graphic artists, software experts, educational psychologists, and others.

A program aiming to move beyond the present stagnant situation must have three main aims:

(1) To create really first-rate multimedia computerized learning environments;

(2) To define and construct much improved courseware production tools;

(3) To produce a number of model lessons which set new standards for top-quality courseware.

Although progress on point (2) will undoubtedly be slowed by the need for serious technical study and by complex implementation problems, choice of just the right project would allow rapid dramatic progress toward (1) and (3), at least with a few "showcase" examples.

In the next section of this article there are several scenarios illustrating some of the ways in which a graphically exciting learning environment can be used. In reading these scenarios, certain points should be kept in mind. Presently, most CAI courseware designers are attempting to build highly constrained, miniature systems which can run on today's computers, and even on tiny personal computers like the Apple or the IBM PC. Even though the commercial pressures that lead to this concentration are clear, emphasis on today's machines is a very poor way of exploring the true potential of CAI technology. To prepare for the possibilities that the next few years will open up, we need to "lead" the rapidly moving hardware "target" substantially, by writing software and creating courseware not for today's machines, but for systems that will be typical five years or a decade from now. We also need to take courseware very seriously and strive to create "classics" that can be used for years into the future. Finally, we need to design CAI software systems that emphasize machine independence and transportability in a way that will allow courseware written today to carry forward to tomorrow's superior hardware.

If all this is not done, the courseware written during the next few years will be obsolete as soon as it is completed. Worse, it will conceal rather than illuminate the very attractive possibilities which it should be our aim to explore.

What a First-Class CAI Environment Should Provide

A first-class CAI environment should be able to support all of the following interactive scenarios:

Scenario One: A Lesson in Ancient History.

The student, in a college-level world history course, is reviewing the sequence of political events surrounding the collapse of the Roman Republic and the rise of the Principiate of Augustus. Some of the material is presented as straight text, which appears on the screen in an attractive book-like format, with single-frame illustrations, obtained from a videodisc storing up to 50,000 images, occupying roughly a quarter of the screen. This text material is enlivened by occasional (optional) quotations from ancient sources, for example, Plutarch's *Lives of the Noble Romans*, presented in audio, quotations from each such source being read in the voice of a single actor who "represents" the ancient author. This material is announced on the screen by a line reading something like: "Concerning Crassus, Plutarch said the following... ." This line appears in a color which the student recognizes as meaning that an optional audio insert is available. If he touches this line on the screen, the video illustration on the page switches to show a bust of Crassus, while Plutarch's remark on Crassus is read in a voice emphasizing its drama and irony. At later point in the lesson, information concerning the Battle of Phillippi is presented, first by text explaining its significance, date, etc., but then as 5-10 minutes of filmed material taken from the videodisc, and introduced by compelling musical fanfare. A map of Italy and Greece, showing the lines of march of the Republican and Antonian armies to the battle, is displayed. This is followed by a more detailed map of the area of northern Greece in which the battle was fought; then by film footage, taken from a helicopter, of the terrain of Phillippi. After this, a brief enactment of the battle itself is filmed, with views of its commanders in action, voice-over commentary on its main incidents, and dramatic scenes of the suicides of Cassius and Brutus. These file-clips then yield to text reviewing the significance of the battle and the events which followed it, with audio statements from ancient authorities reflecting on its meaning for the Roman world.

This didactic presentation is followed by an interactive multiple-choice quiz on the material of the Lesson. Questions answered wrongly trigger re-runs of relevant portions of text, of audio, and of captioned video stills. In addition, the student has the ability to make abbreviated extracts of the text, audio, and video material presented, so as to build up his own personalized course summary for end-of-term review.

Scenario Two: An Arithmetic Lesson.

Here we envisage a grade-school student trying to master "long" addition (i.e., addition of multi-digit numbers). The student is assumed to have memorized the single-digit addition table (however, the lesson software remains alert for errors in single-digit arithmetic, and will review items of this material, using appropriate displays and audio statements, when appropriate.) It is also assumed that, in a prior lesson, the student has been given practice in distinguishing sums like $6 + 5$, which yield a result of 10 or more, from sums like $4 + 3$ which are less than 10. Finally, it is assumed that the idea of "carrying" basic to multi-digit addition has just been presented in a classroom lesson.

The computerized lesson begins with a brief video clip of a teacher reviewing this basic idea at a blackboard; the teacher works several sums, points to relevant features of the numbers being added, writes carries on the board, emphasizes the right-to-left nature of the addition process, etc. Then the student enters an interactive review and practice sequence covering the material just presented. A practice sum, for example,

$$\begin{array}{r} 375 \\ + \ 419 \\ \hline \end{array}$$

appears on the screen. A voice prompt, telling the student that he is to participate in adding these numbers, is given. However, the student is only required to handle the carries; the computer will handle the single digit additions for him automatically. As the turn of each column to be added arrives, the column is highlighted by color, blinking, and a graphic arrow. A two-item selection menu, offering a choice of "carry" and "don't carry" appears on the screen. The student is expected to touch the appropriate one, and when he does so, the computer will write the associated sum-digit in the appropriate column and advance to the next column. Any error will trigger appropriate video and audio diagnostics, and a short review of earlier material.

A first illustrative sum is run through in completely automatic fashion, with a graphic hand representing the student's finger, and a running audio commentary explaining what is expected at each point. Then several interactive examples requiring student response are run. If these are handled successfully, the student is offered an optional video game which tests the speed with which he can handle problems of this sort. In this game, moving spaceships bearing long sums appear on the screen, under pursuit. The student continues to interact by touching the carry/do not-carry menu. Incorrect responses cause the spaceships to explode, losing a point; correct responses allow them to evade pursuit. Delayed responses cause the spaceships to slow down, making them subject to demolition by their pursuers.

After this lesson subsegment, drill resumes. Now the computer handles the carries, and the student is required to supply the associated sum-digit. This is done by touching a 10-item menu of digits on the screen. Next, the student is asked to handle both sums and carries together, and then finally to handle digit-by-digit long addition without any computer assistance. Errors indicating problems in handling either carries or single-digit addition trigger reviews of prior material. Review games utilizing all skill levels are available.

Scenario Three: A Geometry Lesson.

The arithmetic-lesson scenario just presented illustrates several principles important to courseware design.

(a) Ways must be found to focus students' attention on the

skills which are being taught, not on ancillary details of the computer environment. For example, use of a typewriter keyboard distracts attention from material on the screen, and should be avoided whenever possible. Hence the use of touch-sensitive menus displayed directly on the screen. Similarly, having to read written instructions detracts from focus on other screen items, and should therefore be avoided when spoken instructions or direct graphic emphases can be used.

(b) Many conventional academic skills simply amount to the ability to select and apply algorithmic or near-algorithmic procedures rapidly and correctly. These skills are built up from subskills, e.g. to add long digits one must know that the process moves from right to left, must handle single digit sums correctly, and must know how to manipulate carries. Courseware can concentrate on one subskill at a time, in a manner impossible for a textbook and hardly available to the classroom teacher, namely by asking the student to handle only that part of a procedure on which pedagogical stress is to be laid, while other aspects of the same procedure are handled automatically by the computer.

The geometry-lesson scenario which we now present emphasizes this latter point, but in a more sophisticated way than the simpler arithmetic lesson. It also shows the extent to which CAI lessons will have to be backed up by complex symbol-manipulating software. We envision a high-school level student working his way through a conventional course in plane geometry which emphasizes proofs and the solution of "original" problems. The elementary basis of many proofs of the kind is the systematic use of triangle congruence to prove equality of various angles and distances in a given figure. A computer can easily apply these rules systematically and rapidly: a proof can therefore be considered entirely "elementary" if it can be given using this method repeatedly, and if no other idea is required. However, if this one tool is insufficient, a less elementary proof must be given. Often the key to such a proof is to add a few lines or curves (e.g. circles) to the original figure: the proof is then elementary in the expanded figure.

An interactive lesson can exploit this fact, which is crucial to conceptual mastery of high-school geometry. For this, a succession of problems requiring proof, each accompanied by a figure presented on the computer's display screen, can be presented. The computer can then reason as far as possible using congruent triangles only. It can list its conclusions, and, if the student so requests, can review the steps of inference which led to each conclusion. (Such review will emphasize the figure elements involved in each step graphically, and will use audio commentary. If requested by the student, statements of any prior theorems used in the computer's inferences will be retrieved and displayed or spoken.) However, some problems will be too difficult to yield to such elementary reasoning. In these cases the student will have to discover the line (or lines or curves) which will make elementary proof possible when added to the figure. As a hint, the computer offers a multi-item touch-sensitive menu: "Shall I add a line?" "Shall I add a circle?" Suppose, for example, that the student chooses to add a line. Then a menu of possible lines will be offered: "Line between two points," "angle bisector," "line perpendicular to specified line through specified point." When the student chooses one of these by touching it, labels will appear in the figure being analyzed, and then by touching these labels the student can signify exactly what new item he wants added to the figure; these elements will appear immediately in the figure. (They can be erased just as immediately if the student decides that they are not what he wants.) Once these additions are complete, the computer will apply elementary reasoning-by-congruence to the expanded figure. If the right lines or curves have been added, the proof will now succeed; if not, the student must continue wrestling with the problem.

This scheme, which combines student interaction with sophisticated computer assistance, has the merit of focusing attention on the key strategic and conceptual decisions needed to handle a problem, while lower-level, assumed skills are automatically supplied. This recipe is applicable to many other situations in mathematics and science. For example, a student learning to deal with word problems in algebra may only be required to read English-language problem statements and convert them to algebraic equations; the computer can then solve these algebraic equations automatically. Similarly, a calculus student learning how to evaluate indefinite integrals can be offered a menu of standard manipulations: "integrate by parts," "substitute for a variable," "decompose the integral as a sum," "add and subtract a term to the integrand," "factor," etc. Once an alternative has been chosen and its parameters supplied, the necessary manipulations can be performed automatically. As already said, by relieving the student of part of the burden of low-level arithmetic and algebraic details, we focus her attention on the conceptual content of the material on hand, thus allowing her to comprehend it more richly. This should be of significance both to the strong student, whose progress the computer can accelerate, and the weaker student, for whose inaccuracy in applying the auxiliary manipulations which support higher level skills the computer can compensate. Hopefully, as weaker students come to understand how their skill deficiencies impede their ability to handle interesting problems, and as intensive interactive experience pinpoints these deficiencies, they will be motivated to overcome their own shortcomings by systematic review and practice.

Many other equally compelling scenarios illustrating the combined use of video, sound, and computer graphics could be presented. In courses such as physics or chemistry which traditionally involve laboratory demonstrations, video clips of demonstration experiments, highlighted appropriately by dynamically evolving graphs and charts, can be presented. Significant aspects of a demonstration can be emphasized using audio, e.g. the notion of acceleration in elementary physics emphasized by emitting clicks at a rate proportional to the velocity of an accelerating body. Quizzes on the significance of an experiment can produce specially highlighted video reruns of key moments in the experiment when wrongly answered. Experiments in genetics or biology can be presented in speeded-up video representations to emphasize principles of growth. Video clips can display a much larger variety of experimental equipment than could conveniently be shown in a university laboratory.

The very large information storage capability of videodisc can be used to hold various specialized forms of "encyclopedias" with which the student can interact. For example, a student in an organic chemistry course could be asked to design molecules, which he specifies by connecting carbon, hydrogen, nitrogen, and other atoms in various patterns using a graphic editor. The actual properties of the molecules he suggests can then be retrieved from an on-disc chemical encyclopedia and compared to the desired properties. Various 3-D representations of these molecules could also be displayed, allowing the student to deepen his intuitive understanding of their chemistry.

Similar uses of videodisc for social science training are possible. For example, extended psychotherapeutic interviews with a series of cooperating patients can be decomposed into question-response pairs and stored on videodisc. Student therapists could then be presented with a menu of possible interview questions and asked to decide what sort of response they would expect a given patient to make, following which the patient's actual response can be retrieved and played back.

The CAI environment assumed in the three preceding scenarios combines graphics, voice output, music, video, major textual and image data bases, and very sophisticated software. It is now time to say something about the cost of all this. We will first review costs associated with the necessary hardware, where current trends are strongly favorable, and then discuss the deeper and more vexing question of software costs.

Costs

It is clear that the rate at which CAI systems penetrate the schools will depend not only on their attractiveness but also on their cost. For this reason it is appropriate to examine present trends affecting the costs of the major components of a first-rate learning environment. The principal "hardware" elements of future CAI sytems are likely to be as follows:

(A) *Processors.* For these, the basic "computers" which will supply basic computational power to the CAI systems we envisage, the situation is very favorable. The power of a computer is conventionally measured in standard "MIPS" (millions of arithmetic instructions per second). Typical current (early 1983) inexpensive personal computers supply about 1/10 MIP of computing power. The new generation of 32-bit microprocessors, like the MOTOROLA 68000 processor used in the newly announced APPLE Corporation "LISA" system, rate at about 1/2 MIP. By the end of 1983, this chip should be available in an upgraded form running twice as fast. Other microprocessor chips, now in earlier stages of development, should attain computing rates of 5-10 MIPS by the end of the decade. This should make $100 computing engines, roughly as powerful as present large scientific computers, broadly available by 1990.

(B) *High-speed memory.* In present technology, the data which computers manipulate, select, rearrange, and display is stored on media having one of two radically different characteristics. Data items, for example, numbers used in internal calculations to produce graphic output, which must be available within one or two high-speed computational cycles, i.e. within a millionth of a second or less, are stored on integrated-circuit "memory chips." Other data, for which access delays of 1/10 second are tolerable, is held on rotating storage discs, which resemble ordinary long-playing records but which rotate faster and on which information is recorded magnetically and hence erasable.

The amount of high-speed memory available in a computing system affects its performance significantly. If all of the programs and data needed by the computer can be held in high-speed electronic memory, programming the computer is simplified considerably. If, in contrast, high-speed memory is in short supply, programming is seriously complicated by the need to use more intricate, compressed representations of necessary data, and by the need to move fragments of program and data back and forth between high-speed and bulk memory. Faced with this technical challenge, programmers will cut back on the level of external function which they attempt to supply, and increased programming costs will limit the range of software products which become available. The result seen by the end-user will be systems which are either less sophisticated, less lively, or both. Hence the cost of high-speech memory, and the amount of it available for program/data storage, is an important factor determining the adequacy of hardware for CAI use.

A second important use of high-speed memory in the CAI context is in the "refresh" memories which store images displayed on a graphic screen. These images are built up as matrices of individual dots ("pixels") of specified intensity and color. A 512 by 512 matrix of dots gives a high quality image, and a 1000 by 1000 dot matrix gives an excellent, almost film-quality image. If one character or "byte" (equivalent to two decimal digits) is available per image pixel, half of this data can be used to select one of 16 intensity levels, and the other half to select one of 16 colors. If two bytes are available per pixel, 16 intensity levels and roughly 4,800 colors become possible. Thus, a graphics screen backed up by 2 million bytes (characters) of data will provide excellent, glossy-magazine quality graphics. If this allotment of memory is doubled, then while a prior image is being displayed, the computer can be writing the next image to be viewed into a behind-the-scenes image memory. Switching between these two images then becomes effectively instantaneous, eliminating the psychological irritation caused by visible manipulation of images while they are being viewed.

At present, the manufacting cost of high-speed memory is about $1,000 per megabyte (one million characters of storage capacity.) This is based on the use of memory chips storing 8,000 characters each, which are currently available for about 3 or 4 dollars. Improved chips, storing 32,000 characters each, are about to appear on the market. Before 1990, these should be available at about the present price of the 8,000 character chips, so that high-speed memory should be available at about $250 per megabyte manufacturing cost. The 1990 manufacturing cost of 4 million bytes of memory for the very luxurious graphics environment described above, plus an additional 4 megabytes of memory for program and data storage, can therefore be estimated as roughly $2,000.

(C) *Bulk data storage.* The bulk data required by an interactive CAI system falls into two principal categories, fixed and variable. Fixed data comprises that mass of recorded text, video, and audio material which the computer driving a CAI system needs to retrieve and rearrange, but not to store in bulk after dynamic rearrangement. The medium of choice for all this material is digital videodisc.

The second bulk data category, variable bulk data, includes student-collected text excerpts, massive intermediate results generated by computer calculations (e.g. "workspaces" of temporarily suspended program runs or simulations), screensful of graphic material held temporarily for later redisplay, dynamically generated audio material held for subitem selection or replay, etc. This material is best stored on rotating magnetic discs.

High quality magnetic storage discs able to hold 20-30 million characters of recorded information presently retail for about five thousand dollars. Improvements in recording techniques can be expected to increase this capacity significantly. By the end of the decade, 200 to 300 million character discs in the $1,000 to $2,000 price range should be available. Some technological forecasters in this area are predicting even greater improvements in storage density, based on the use of improved miniature recording heads produced using integrated circuit techniques that might possibly increase the amount of information stored on a disc by factors as large as 100. This will provide variable bulk data storage capacity adequate to CAI systems requirements.

Videodisc already provides an enormous capability for storing fixed data. Laser videodiscs of the kind recently introduced by Phillips, MCA, Sony, Pioneer, and various other manufacturers store roughly 10 billion characters of information, equivalent to roughly five million pages of printed information, 200 hours of high quality audio, or 50,000 video frames (1 hour of live video). Technological improvements over the next decade may increase these already very great capacities by a factor of 10. The cost of individual disc manufacture is somewhere around $5, and quality computer-controlled players are already available for between five hundred and one thousand dollars.

Future CAI material will in all likelihood be stored on video-discs, several lessons being stored on each disc. The principal constraint on the number of lessons held on a given disc is clearly the number of minutes of live video required: storage require-ments for single images, audio material, printed text, and com-puter programs are trivial in comparison. Hence it is possible to store large amounts of supplementary reference material on a disc, provided that only still images, without moving video, are used in these supplementary sections. For example, a student in our hypothetical world history course can be offered full texts of all the ancient authors quoted, biographies and portraits of the historical figures cited, additional maps, pictures of Roman military equipment, etc., all of which can easily be stored on the disc which holds his history lesson. As storage capacity rises toward 10 hours of video per disc, it will be possible to store entire courses on no more than a dozen discs.

(D) *Display terminals.* High quality color display terminals are currently available for $400 to $600. This is already a mass-production price, and cannot be expected to fall significantly over the period we consider. Terminal prices should be largely independent of the resolution provided (given that we count the principal cost associated with higher resolution, namely the larger refresh memory required, separately). Touch sensitivity is not at present a standard terminal feature, but it can easily be provided, e.g. by installing rows of light-emitting diodes and matching receptors around the edges of a terminal's screen, and should add little to the quoted price of a terminal.

(E) *Audio output.* A CAI system can make effective use of two principal forms of audio output:

(1) *Pre-recorded music and speech.* As much of this material as is likely to be needed can be held on videodisc and selected for playback under computer control.

(2) *Computer-generated speech.* Speech of this kind is generated by transmitting text, possibly written in some special phonetic alphabet, to a speech-generating device which expands it into a standard audio signal. This involves application of some form

219

of "pronunciation algorithm," which uses rules of context to fill in details of the audio signal pronounced. This level of detail includes customary elisions and transition patterns which are not explicit in the text supplied. To be done well, such "phonetic expansion" requires a small data base representing various "exception rules" of English pronunciation. A high-quality speech generator will also handle inflections, allow emphasis, etc. A speech generator of this quality would currently cost several thousand dollars, but it should not be hard to reduce the signal generator and other circuitry required for high-quality computer speech to a small number of chips that can be manufactured for well under $100.

The technology used for speech generation should also support production of computer-generated music, making elaborate orchestral effects available to the courseware designer at little or no additional cost.

We summarize the preceding discussion of component prices in the following table of hypothetical 1990 costs, to a school system buying in bulk, of the high-quality multimedia CAI system that technology will support. Anticipated manufacturing costs noted above are multiplied by a factor of 3 to obtain estimated prices. (This is an unrealistically low markup for the present computer industry, but quite reasonable for the future mass-market operation which this discussion assumes.)

Estimated Multi-Media CAI System Hardware Component Prices (1990)

Processing element (10 MIP microcomputer)	$ 300.
4 Megabytes high-speed storage	3,000.
300 Megabyte magnetic storage disc	1,000.
Videodisc player, computer controlled, with text & multi-channel audio capabilities	500.
Touch-sensitive display screen with graphics and color video capabilities	400.
Loudspeakers	200.
Computerized speech/music generator with high-quality pronunciation algorithm and inflection capabilities	150.
Supplementary electronic equipment (including network interfaces)	200.
Estimated total costs	$5,700.

A system of this type, with a somewhat less powerful processor, can be assembled today for about 10 times this estimated cost.

CAI Languages for Creating Quality Courseware

CAI languages (or coursewriting languages), are the technical tools used to compose computer-controlled interactive lessons for display on terminals furnished with a variety of supplemental input and output devices. As the conditions of computational resource scarcity that constrained yesterday's CAI systems relax, the design and software problems which must be solved to create educationally effective courseware become increasingly central. However, the software tool kits provided by current coursewriting languages are still quite modest, and a substantial experimentally guided tool-building effort is needed if these tools are to be made adequate to their intended task.

The success of a CAI system will be strongly conditioned by the adequacy of the coursewriting language it makes available. A good language will aid in the creation of lively, flexible pedagogical dialogs which succeed in capturing and holding student attention: a poor language will make liveliness hard to attain, and conduce to stilted, book-like lessons which destroy student interest and motivation, and which wind up teaching computer-related details and idiosyncracies rather than their ostensible subject matter.

Moreover, as we have seen, creation of high-quality courseware will be expensive; the CAI language used to write courseware is the principal tool available for controlling this expense. To amortize the high cost of courseware, we will need to run it on a variety of computers; thus the courseware must be portable from machine to machine. This implies that the CAI language used to write the courseware will also have to be highly portable. But even the most careful attention to transportability will eventually be undercut by the major changes in hardware now under way, and by the development of radically new output devices. Hence we must expect rewrite of existing courseware to become necessary from time to time. If the major human effort involved in the creation of high quality lessons is to carry forward past such rewrite, it is also important that courseware be *well-documented* and *readable* enough for its essential structure to be easily visible.

To secure all these advantages, a powerful, transportable, standardized, very high-level CAI language is required. Even though current CAI languages suggest features desirable for such a language, none of them are well-organized or transportable enough to serve as such a language without very major changes. Experimental work on such a tool, pursued in an environment providing first-rate hardware, should be regarded as a main developmental priority over the next few years.

The following paragraphs comment on some of the facilities which such a language will have to support.

(A) *Powerful general-purpose computational facilities.* Courseware will occasionally need to make calculations of the sort supported by general-purpose algorithmic languages, for which reason an authoring language should make all the facilities of a powerful algorithmic language available. These should include flexible string manipulation and analysis operators, pattern-analysis mechanisms, operators facilitating use of general tables and maps, and a capability for handling very general *composite objects*, for example, internal representations of the shapes of complex geometric bodies. New types of objects should be definable by the courseware writer, and the meaning of operator signs should be conditioned by the types of the objects to which the operators are applied. Since some of the operations used in CAI languages will have numerous inputs, many of which remain constant over several successive invocations of the operation, partial pre-processing of these inputs is a useful facility. Since simulation-based lessons can be expected to be common, simulation-oriented parallelism, i.e. the ability to define various semi-independent event streams that the computer can push forward more or less simultaneously is a semantic facility that is important to provide. Finally, since it is so powerful a tool for response analysis, generalized backtracking, i.e. the ability to carry out computational activities provisionally but to eliminate all their effects when and if they prove fruitless, is a desirable semantic mechanism.

No existing CAI language supports all the facilities listed in the preceding paragraph at all fully, even though a modest general-purpose computational capability is available in all CAI languages.

(B) *Input facilities.* In addition, an effective CAI language should support input from a touch-sensitive screen, and from various kinds of "tablets," "joysticks," and "mice," which can be used to pass graphic information to a computer. These facilities should support various useful forms of "rubber banding," i.e. for drawing lines and other figures which stretch and shrink dynamically as a student's finger moves on a touch-sensitive screen.

(C) *Output facilities, including graphic output.* To attain the degree of liveliness that can ensure student attention, a CAI language should support a variety of output facilities, including:

(i) two-dimensional text output, in a variety of alphabets, fonts, letter sizes, and character orientations;

(ii) two-dimensional graphic output;

(iii) two-dimensional graphic representation of three-dimensional figures;

(iv) animation, and in particular multiple animated "sprites" of the kind used by video-game designers;

(v) video output in still-frame and continuous-play modes;

(vi) electronically generated audio-signals;

(vii) high-quality computer-generated speech;

(viii) audio playback of narration and music;

Systematic 2-D graphic output facilities can be provided by incorporating a few powerful 2-D graphic primitives in a standard CAI language. The following facilities are desirable:

(a) 2-D objects, made up of points, line segments, circular arcs, and other smooth curves should be manipulable by standard geometric transformations, including translation, rotation, and expansion. Simple geometric objects should be combinable into composite objects which can be manipulated as wholes.

(b) 2-D transformations of general mathematical form should be definable, and an output primitive which draws the image of a 2-D object under such a transformation should be provided.

It should be possible to draw objects using either solid lines or lines dashed in a variety of patterns. This will give the course-ware author the ability to create various types of systematically distorted figures, making various interesting mathematical and graphic effects possible.

The development of high-power graphic chips may make it feasible to use very high quality 3-D "synthavision" images.

(D) *Response analysis.* Analysis of student responses is an area of great importance. Student responses typed at a terminal can be regarded either as character strings or as sequences of words. For example, the Illinois TUTOR language supports both views of student input, and also facilitates acquisition and analysis of algebraic and numerical expressions entered by a student. TUTOR's string-analysis features are a narrow subset of those provided by SNOBOL, and could well be replaced by something much closer to SNOBOL's pattern match facility. However, TUTOR's word-oriented match facilities embody other interesting ideas. When these facilities are applied to a string S, S is regarded as a sequence of words (separated by blanks and punctuation marks), which are matched to a specified word pattern. Unless exact match is specified, words will match if a standard spelling-error algorithm judges them to resemble each other closely enough to be regarded as misspellings of each other. A word pattern consists of a specified sequence Q of words, together with a set of "ignorable" words which can appear in the student input but will be ignored by the matching process. The individual items in the sequence Q are actually sets of logically equivalent words, any one of which is acceptable in a given position. The match operation can specify either that all words in the student input must appear in the pattern to which the input is matched (full match) or that a subset of the input must match the pattern (subset match.) It is also possible to specify either ordered match (in which case words must appear in the student input in the same order in which they appear in the pattern being matched) or unordered match (in which case the elements of the pattern can be permuted arbitrarily without disturbing the match.)

All this is good as far as it goes, but deeper methods are urgently required. Current research in computer-analysis of natural language is steadily extending the ability of computer programs to deal with unrestricted natural language input. However, this ability is still highly imperfect, and may not be usable at all until it becomes entirely solid. Even if a computer handles 90% of a student's natural language inputs adequately, bizarre reactions to 10% of these inputs, or even inability to produce any reaction, may create an unnerving pedagogical environment, which forces the CAI designer to prefer multiple-choice inputs chosen from a pre-restricted menu. For this reason, and also because of the presently very high computational cost of natural language input, we can expect the use of such input in courseware to develop slowly.

On the other hand, it may be possible to develop inexpensive speech-recognition chips which can deal adequately with very limited classes of utterances, e.g. "yes," "no," the numbers and letters, etc. This could be a valuable adjunct to touch-sensitive screen input, and should be supported by a coursewriting language if and when it becomes available.

It is worth emphasizing that the preceding discussion of CAI software tools is deliberately humdrum. It emphasizes software facilities that can be developed now, and by intent ignores many exciting software and courseware possibilities having too large a research component for their immediate construction to be straightforward. To redress this perhaps overly pragmatic attitude, it is worth describing a few of these more speculative possibilities. Some of these relate to our steadily improving, although still imperfect, ability to process natural language. For example, an English-language parser might work with a grammar comprehending both the constructions of standard English and those of other less "literary" language versions which students tend to use. Sentences in compositions which were not quite grammatical could then be parsed, and trigger either review of the grammatical rules being violated, or paraphrases of the offending sentences which cast them into literary English. (A parser of this sophistication could also work out sentence diagrams for the student's edification.) A somewhat more sophisticated parser might also be capable of detecting common faults of style, e.g. the use of passive constructions where shorter active phrasings are available, and might be able to generate various improved reformulations of a user's sentences for his inspection and possible adoption. Though it would have to operate at a much more sophisticated syntactic level, such a program could be regarded as a syntactic analog of the spelling-error detection programs that have now become common.

A wave of enthusiastic attention is now focusing on so-called expert consultant systems, which attempt to diagnose real-world situations, assigning each such situation to one of a finite number of diagnostic classifications by asking sensibly ordered sequences of questions. It is hoped that systems of this sort will be capable of imitating the investigative model used by human experts closely enough to approximate expert behavior. If this hope proves justified, it may also prove possible to use these systems to compare expert behavior with novice behavior, and, e.g. to automatically critique students in analytical chemistry or medical diagnosis classes for asking questions that are probably irrelevant, or for attempting laboratory manipulations which an expert would reject as being unlikely to succeed. These intriguing possibilities relate to the broader question of the extent to which CAI systems can develop valid models of a student's skill inventory and cognitive biases. To do this will plainly not be easy, but to the extent that such an attempt succeeds, CAI systems will come to behave like sophisticated, infinitely patient tutors.

On Composing Quality Courseware

It should be obvious from the scenarios presented in section 2 of this article that composition of quality software involves the combination of many skills. To choose the points of emphasis and the key pedagogical approaches around which a lesson or course will revolve clearly requires a scholar deeply versed in the subject to be presented, who must also be a talented and sensitive teacher, aware of many possible lines of presentation, and who is able to select the presentation best adapted to the CAI medium. These academic skills must be combined with the skills of graphic designers, with the image-oriented, "filmic" skills of the motion picture director, and with much programming expertise, ranging from a knowledge of symbolic manipulation algorithms to video game design. Actors able to read passages of text with dramatic emphasis will also be needed, and film clips illustrating significant moments in history and recording important scientific experiments will have to be either collected or produced. In some cases, elaborate graphic animations will be the most effective way of making a pedagogical point, so the services of animation studios will be needed.

It is clear that production of material of this complexity will resemble the activities of a television or Hollywood film studio more than it resembles text-book production, a cottage industry by comparison. The kind of person around whom such production will revolve will be a director-like figure, i.e. a talented coordinator able to think in media terms, rather than an academic. Subject-area scholars will play something like the role that writers play in Hollywood, namely they will generate skeletal, first phase conceptual structures, whose realization will then require much larger groups and different abilities. The organization of programmers, animators, and cinema talents which Lucasfilms has built up to carry forward their "Star Wars" series of films shows approximately what will be needed.

It is clear that production of material of this complexity will be quite expensive. Five million dollars per hour is regarded as a modest budget for production of quality film. The possibility of producing a film of "Star Wars" quality depends on the fact that it can earn tens of millions of dollars within a month of its release. Where is the money to support comparable costs of first-rate CAI courseware to come from?

The students in a given academic grade constitute an audience of roughly three million. With proper organization, which would have to involve not only producers of educational materials, but also local school boards, plus state and federal educational authorities, it might be possible to extract $5/day per pupil, i.e. roughly $1,000/year from the school system for the production of educational courseware. This is $3 billion per year per grade level, with which hardware must be acquired and roughly 1,000 hours of courseware must be produced. This budget is ample, but will only materialize if use of CAI materials in schools becomes ubiquitous, and if a single system able to run all lessons becomes a nationwide standard. Such a high degree of national coordination failing, the revenue available to support courseware production will be divided and drastically diminished, lowering the quality of the lessons which it is possible to produce.

This difficulty will be most severe in the first phases of CAI system development. The present educational market for computers is dominated by inexpensive machines of very modest capability, hardly resembling the deluxe future teaching environments envisaged above. Systems are unstandardized, especially in regard to essential graphic capabilities. Software designers wishing to secure a large market must write for a least common denominator of hardware capabilities, for very restricted memories, and in primitive languages like BASIC apt to be available on all machines. The mediocre products which come forth in these circumstances do not tempt schools to any major commitment, or inspire any larger view of what might be possible. Publishers are skeptical of the revenues obtainable from so fragmented a market.

To begin breaking this interlocking log-jam of difficulties, it is appropriate for the Federal government, perhaps acting in conjunction with a few state governments and private foundations, to sponsor the development of a few model CAI environments, and of a limited number of first-class computerized courses. While it will be extremely difficult to produce even a partial set of materials right now, this might provide a setting to which publishers and school officials could be exposed, thereby heightening awareness of the potential of CAI and the extent to which all the main technological items which it will require are already at hand. Moreover, once suitable equipment becomes available and a few CAI courseware "studios" come into being, it may prove possible to obtain substantial initial revenues from organizations (e.g. the military, large corporations with substantial employee training activities) able to pay high initial prices for courseware. In this way, the infant courseware industry might take the first steps toward realizing its potential.

A Final Comment

Developers of truly effective computerized instruction systems will draw upon many sources of ideas, ranging from recent developments in cognitive science to the informal but essential wisdom of the skilled teacher. Technological developments in fields other than VLSI chip fabrication will contribute to their opportunity: for example, falling communication costs will make it easy to tie separate CAI workstations together, thus making it possible for geographically separated students with like interests and abilities to share insights and efforts while they study together. However, the core of the present opportunity is technological and developmental: to build a hardware/software environment within which attractive, exciting, carefully directed, smoothly flowing, and richly informative material can be brought to the individual learner; to create those few demonstration lessons and courses out of which a broader commercial activity can grow; and to assess/refine this experimental material by observing the way in which test groups of students interact with it. From this perspective, the most compelling initial step is clearly to organize several CAI studios within which multimedia educational workstations supported by software systems which facilitate their use become available. The core staff at such an institution should aim to provide first-class hardware, software, graphics, and image-processing support to teams of subject matter specialists, experienced teachers, and visual artists who approach it with creative ideas for composing individual lessons or groups of lessons. Though in time psychologists and educational statisticians will undoubtedly contribute to the work of such a studio, whose activities they are sure to find worthy of study and analysis, it would seem inappropriate to tie the work of such a studio too closely to any specific psychological theory, rather than organizing it as a neutral locus for the experimental development of CAI materials.

1983

Chapter Twenty

Misreading the History of Mathematics

THE HISTORY OF MATHEMATICS, like most fields not supported by Federal Funds, is a disaster area.

The few practitioners can be roughly divided into three classes. At the upper end, we find the Great Man — two or three of them living at a given time — who, with commendable *Gründlichkeit*, sets out at age eighteen to write the History from its beginning. Unfortunately, he dies at age 92 after completing the Egyptians, and leaving only a rough draft on the Cretans. Throughout his life, he remained blissfully unaware that anything that happened before Leibniz is not history, but paleontology.

In the middle we find the Pedagogue, modestly satisfied with producing a solid textbook and occasionally correcting one crass prejudice of many that are now plaguing the scientific masses. What a distance from the outpour and quality of historical studies on Shakespeare!

At the lower end we meet the Amateur. Gliding with hare-brained frivolity across places and periods, he (or she) comes up with a concoction of anecdotes, sensationalism, and half-truths, avidly read by students and high school teachers, and invariably achieving the crowning recognition of the paperback edition.

In bygone days, chivalry would have prevented us from classifying Mrs. Reid;[1] liberated by recent social changes, we can

---

[1] Constance Reid, *Hilbert,* Springer Verlag.

treat her work like that of any colleague. Although she squarely belongs at the lower end, we hasten to add that her book contains no half-truths whatsoever. It is factually impeccable and intellectually a fiasco.

Many of us remember Old Europe and its *mitteleuropäische Kultur*. The sleek blue express trains would pull out from Milan and Rome, Venice and Florence, bearing the tag *Mitropa,* carrying away haughty ladies with gutteral accents, who would get busy on their manuscripts as soon as the train set in motion. Their heavenly destinations, Vienna and Berlin, Prague and Stockholm, were pictured as the meccas of exquisite artistic refinement and unfathomable learning, a vision shared with proud superiority by the natives of *Mitropa.* To them, Americans were blundering infants, Italians frivolous fakers, Englishmen eccentric gentlemen who cultivated Greek or mathematics as a hobby, everyone else a barbarian. Only the French could stand up to them, a painful thorn in the thick layers of their cultural fat, a feared enemy richly endowed with the deadly weapons of wit and elegance.

It is hard to believe that this very same world of stifling *Kultur* and deadening prejudice (now largely wiped out) produced a majority of the great minds of the West, among them David Hilbert.

If Mrs. Reid were a historian, she might have begun her biography with a description of this background and its interaction with Hilbert's intellectual development. Instead, she begins it with an excursion into genetic mythology. Following the equivocal party line of those who arrogantly believe that human genetic behavior can be inferred—by an irresponsible extrapolating—from our knowledge of the genetics of fruitflies, she conscientiously lists ancestors and their trades, in a vain search for the forebodings of mathematical talent.

What if she succeeded? Would her findings help us understand the *person* David Hilbert? She ought to read what a serious biographer writes: "To show that…freedom alone accounts for the whole person; to display freedom in its struggle with fatality, crushed at first by the inevitable but then overcoming it; to show that genius is not given at the start, but the brilliant invention of someone who is looking for a way out; to reconstruct the original choice of a meaning of the universe, a style, a pattern of thought, all the way to tastes and mannerisms... [this is the purpose of biography]."

We leave it to the reader to decide where Mrs. Reid's book belongs, judged by these standards. Apparently, she is under the illusion that a man's biography consists of a collection of facts, chronologically strung together and presented in passable English. And facts she dishes out aplenty: from the obvious to the intimate, from the irrelevant to the crucial, all the way to the minutest detail ("He was fascinated by the *Pferdespulwurm*") and to the anecdote-heard-a-thousand-times.

But an inert collection of facts is not a biography, and what little one can learn from Mrs. Reid's presentation has to be inferred by the reader, leaving him — or her — more often than not, in want of explanation. What, for example, are we to make of Hilbert's relationship with Käthe Kollwitz? Was Franz Hilbert mentally retarded, or was his behavior a result of his father's miseducation? Why and how did the depression come about? (Such downturns do not come out of fatigue alone.) When and, most important, how did Hilbert's intellectual activity begin to fade? (It seems this happened fairly early, but Mrs. Reid, awed by the holiness of her subject, does not give particulars.) What are the details of Hilbert's quarrel with Husserl, which led to the philosopher's transfer to Freiburg? (Mrs. Reid withholds this last detail. It seems that Hilbert had no taste for philosophers — see e.g., the remarks on Kant — Hans Rademacher told this reviewer that, after Heidegger once lectured in Göttingen, Hilbert gleefully repeated to everyone the phrase "*Das Nichts nichtet die Nichtung*").

Mrs. Reid wisely avoids all pretense of scientific biography; what little mathematics she describes is superficial. Exceptions in bad taste are the platitudes on invariant theory in Chapter V, and the slurs on Gordan and Study. (It is time someone wrote an accurate biography of invariant theory, explaining the ambiguous sentence "Hilbert killed invariant theory," now repeated with ovine credulity by everyone.) She does, however, attempt some sort of intellectual biography, and here the main questions again remain unanswered. Hilbert's habit of "Nostrification," as Richard Courant used to call it, should be described and understood, bad trait that it is. Mrs. Reid does not even say in so many words that Hilbert's "Foundations of Geometry" is partly cribbed from Kohn, Schur and Wiener. Hilbert's errors — as found and corrected by Olga Taussky-Todd — remain in the realm of curiousity; they might reveal so much. And the in-depth story of his most stunning discoveries — the proof of Waring's conjecture, for instance — is not told. By what process of elimination did Hilbert arrive at the proof? How did his mind work? For contrast, the reader might glance at Littlewood's *A Mathematician's Miscellany* where in a few pages he will learn what makes Littlewood tick. Or he might compare Mrs. Reid's book with some of our best intellectual biographies, such as Dilthey's *Schleiermacher,* Sartre's *Flaubert* or Croce's *Vico.*

We could go on indefinitely asking questions of this kind, which only bring out Mrs. Reid's inadequacies. It is not her fault if the first full-length biography of our greatest mathematicians had to be written by someone who hardly knows how to differentiate and integrate, and who is not a trained historian to boot. It is instead a sad reflection on our antiintellectual age, which discourages scholarship and the spirit of synthesis in favor of the ephemeral novelty of compartmentalized research.

May Mrs. Reid's effort at least stimulate a reform of our mathematical values, and lead to an authentic concept of mathematical history.

1974

CHAPTER TWENTY-ONE

The Wonderful World of Uncle Stan

I MET STAN ULAM IN 1964, at a lecture I was giving in New York. Kac had prevailed on his reluctance to sit for an hour in front of anything but a coffee table in a sidewalk café. After twenty minutes he remembered "an urgent appointment downtown" and walked out. Years later I was to learn that ten, not twenty minutes is his normal limit for a lecture. Once the idea has come across, he reasons, it is more fun to work out the details by yourself; and if the main idea has not been put forward in the first ten minutes, then the lecture is probably not worth your time.

This one-shot quality of Ulam's thought became more apparent after I arrived in Los Alamos. "Any good idea can be stated in fifty words or less," he pronounced with a twinge of challenge in his voice. I am probably the only student of physics to be taught in twenty-word sentences unevenly spaced throughout the working day, and at later dinner parties, each one an irritatingly clear summary of one or more chapters of the *Handbuch der Physik*.

At first I wondered how he could perform miraculous feats of computation in his head, until I discovered he did not need to; he simply estimated magnitudes following an unerring instinct. "Knowing what is big and what is small is more important than knowing how to solve differential equations," he would warn an audience of shocked mathematicians. His most daring guesses, whether proposing a new idea for fusion of defusing a pompous theory, are approximations based on seeing through irrelevancies and inessentials.

The anecdote is Ulam's literary *genre*. His mind is a repository of thousands of stories, tales, jokes, epigrams, remarks, puzzles, tongue-twisters, footnotes, conclusions, slogans, formulas, diagrams, quotations, limericks, summaries, quips, epitaphs, and headlines. In the course of a normal conversation he simply pulls out of his mind the fifty-odd relevant items, and presents them in linear succession. A second-order memory prevents him from repeating himself too often before the same public.

His *Adventures* were written by reading the whole repertoire into the tape recorder, with the omission of a few prurient episodes to be made public at a future date. The rearranging and typing was left to Madame. Later, the author inserted appropriate transitional passages into the first set of galleys. The final product compares favorably in style with *The Three Musketeers,* and like that irresistible novel of our youth, after page one, one cannot put it down until the end.

The kaleidoscopic sequence of exits and entrances of personalities of the last fifty years make the book the *Almanac de Gotha* of twentieth-century science. No other scientist has attempted a first-hand coverage of comparable scope, and we can already foreshadow doctoral dissertations in the history of science — and in psychology — being written about a casual remark in the *Adventures*. Such a student's task will be made easier by the author's refreshing belief in facts, and by his shrewd avoidance of "depth" analysis. People are what they are, good and bad, intelligent and stupid, and their behavior is never traced back to the tottering tenets of Freudian mythology.

The factual account of the *res gestae* (the only eyewitness report to date) is pleasantly interrupted by the author's views on science, its future, and its problems. The unassuming exposition, at times running into aristocratic understatement, may lull the reader into taking these remarks as obvious. It is easy to get used to a display of intelligence. One silently appropriates whatever is said, and on second reading the author seems to repeat the reader's lifelong conclusions. The reader might be well-advised to use one of those hemispherical lenses that were once indispensable for proofreading, so that his eye may dwell on the depth of the author's casually stated ideas.

This is a book about success: about the author's success — luck, he would call it — as an emigrant left penniless by the devestation of Poland, about the success of a group of intellectuals, Americans and emigrés together, on a lonely hill in the New Mexican desert, who changed the face of civilization, about the last unquestioned victory of mind over matter. It is a chronicle of greatness written by one of the last survivors of the species. We reread Ulam's memoirs with a longing for a bygone world where physicists could be, like Fermi, both theoretical and experimental, where mathematicians like Everett would wear out a slide rule in a month, where politicians and generals put their trust in Science. A world we would like to bring back, a faith which this book helps us keep alive.

1978

CHAPTER TWENTY-TWO

Ulam

STAN ULAM RESENTED being labelled an intellectual. He would not even agree to being classified a mathematician. He referred to the published volume of his scientific papers as "a slim collection of poems."

Throughout his life, his style in speaking and writing remained the aphorism, the lapidary definition, the capture of a law of nature between one subject and one predicate. "Whatever is worth saying, can be stated in fifty words or less," he used to admonish us, and to teach us by his example.

Mathematics is a cruel profession. Solving a mathematical problem is for most mathematicians an ardous and lengthy process which may take years, even a lifetime. The final conquest of the truth comes, if ever, inevitably tinged with disillusion, soured by the realization of the ultimate irrelevance of all intellectual endeavor.

For Stan Ulam, this process took place instantaneously, in a flash, unremittingly, day and night, as a condition of his being in the world. He was condemned to seeing the truth of whatever he saw. His word then became the warning of the prophet, the mumbled riddle of the Sybil in her trance. The comforts of illusion were denied to him.

His eyesight followed the bidding of his mind. He could focus on a detail so small as to have been missed by everyone. He could decipher a distant rumbling that no one could yet be aware of. But his blindness for average distances kept him from ever enjoying a rest in the quiet lull of mediocrity.

Worried that we might not be ready to bear the burden of those visions, he solicitously improvised daily entertainments, games into which he prodded us all to be players, flimsy amusements and puzzles he built out of his concern that we be spared, as he could not be, the sight of the naked truth.

What saved him, and what was to immensely benefit science, was his instinct for taking the right step at the right time, a step which he invariably took with a scintillating display of elegance.

The inexorable laws of elegant reasoning, which he faithfully observed, became his allies as he drew out the essentials of a new idea, a gem of the mind that he would casually toss off at the world, always at the right time, when ready to be pursued and developed by others. His ideas have blossomed into theories that now grace the world of science. The measurable cardinals have conquered set theory; his foundations of probability have become bedrock. He invented more than one new stochastic process, starting from the imaginary evidence he alone saw beyond the clumsy array of figures spewed out by the very first computers. The strange recurrences of the dynamical systems he was first to describe and simulate are the key to the new dynamics of today.

Stan Ulam came to physics comparatively late in life. With unerring accuracy, he zeroed onto the one indispensable item in the baggage of the physicist, onto the ability to spot and shake the one essential parameter out of a morass of data. In his work at the Lab, he was the virtuoso who outguessed nature, who could compute physical constants old and new to several decimal places, guided only by an uncanny sense for relative orders of magnitude.

Every day at dawn, when most of New Mexico is asleep, Stan Ulam would sit in his study in Santa Fe and write out cryptic outlines on small pieces of paper, often no larger than postage stamps. Rewritten, reformulated, rebroadcast by others to the four corners of the earth, these notes became the problems in mathematics that set the style for a whole period. To generations of mathematicians, Ulam's problems were the door that led them into the new, to the first sweet taste of discovery.

I wish we could have convinced him that his problems will last longer than he expected them to last, that they are and will be the source of much mathematics that is and will be made, that he will still find them around in a next life, sprinkled in the research papers and in the textbooks of whatever time; to convince him that they will brighten our lives, that they will brighten the lives of those who come after us, like a cascade of stars in the crystal sky of Los Alamos, like the fireworks of the Fourth of July.

1984

CHAPTER TWENTY-THREE

Kant

TOGETHER WITH CROSSWORD PUZZLES, quantum mechanics and abstract painting, biography is a creation of the Twentieth Century. To be sure, biographies long and short were written in all ages, but with a few Chinese exceptions the biographies of the past make uninteresting reading. Not even juicy tidbits like Machiavelli's turn of the screw on Cesare Borgia, or the memoirs of the Marquis de Sade, can any longer hold our attention. Spoiled by Sigmund Freud, we demand mud and dirt. Trained to expect crude revelations of sexual misconduct, of brain malfunctions, of childhood traumas, we look forward to one end: the hapless biographies shall be made to be another one of us, complete with a Jewish Mother and manic-depressive fits.

The Twentieth Century — us, to be precise — has shown little tolerance for the exception. Whoever strays from the fold must be either punished or explained. Unable to punish the dead, we are left with no recourse but to fix them with biographies.

Thus, since the *belle époque,* biographers have reaped the benefits of the public's slothful prurience. It takes time and effort to master the intricacies of synthesis *a priori* and of the categorical imperative; how much simpler and more gratifying to read instead the minutest details of Immanuel Kant's daily life, and to share with our hero a common problem of constipation, or an uncommon weakness for strong coffee!

Miraculously, our maligned century is also endowed with an uncanny ability to develop antidotes for the evils it has brought about. The fad of smutty biography eventually sated the public's thirst for banality, and the idea began to make headway that people's private lives show little variation from each other, and those of great men are just as dull as anyone else's. It didn't take another big step to realize that the analysis of personality based upon the presumed relevance of improbable trivia required previous training in the systematic disregard of established rules of logic and common sense. The classic statement: "Kant became a great philosopher because his mother weaned him too early," can nowadays no longer be justified by any amount of persuasion, and one wonders how only a few years ago assertions of this kind could be swallowed hook, line and sinker, by physicians, philosophers and intelligent people alike.

Thanks to the present disillusion with the intimate, today's biographer is confronted with a clear-cut choice. He may yet choose the time-tested recipe, and produce a carefully documented volume (trivia are always carefully documented) that by making minimal demands upon the reader will provide a few hours of stupefied reading, and stand the test of the bestseller list.

Or else, he may decide to write a biography. He will then come face-to-face with the age-old problem of deciding how a man's life relates to his works. In our jaded age, the solutions proposed in the past are simultaneously available and interchangeably plausible.

(a) A man's life *is* his work. Forget dates, data and details, just look at what the man accomplished. There is nothing much to tell about Kant's private life: let's get on to what he did. *Cada uno es hijo de sus obras*, as Cervantes would put it.

(b) A great man's private life *must* be edifying and worth telling; surely, X's weaknesses and Y's nervous breakdowns have been inaccurately recorded. A man shall be great at everything he does, or else he cannot be great. Witness Kant's private life, exemplary in its regularity and predictability.

(c) There are no great men. When shall we stop falling for

cult figures and prima donnas? A biographer's task is to see through the fakery. Not even Immanuel Kant should be spared; tear him down. His life is no different from that of any other *petit bourgeois* of his time, dull and uneventful.

(d) A man's life is inextricably related to his ideas; a biography shall display the unity of mind and action, of life and work. Many other lives of the time are factually similar to Kant's; what made the difference?

Gulyga's biography[1] succeeds in finessing these problems, no easy feat in a philosopher's biography, notoriously the hardest to write.

A philosophical text cannot be understood without the reader's emotional cooperation, be it only sympathy. *Non intratur in veritatem, nisi per charitatem.* An image of the author's personality is thereby conjured up. Perhaps the reader should hold onto that image, the imaginary one, the one the author intended us to have of him, an image which almost by necessity is at variance with reality. For the private lives of some of the greatest philosophers are in pathetic contrast to their writings. Plato bowed to petty tyrants, Leibniz was a con man of sorts, Hegel had illegitimate children, etc. The personalities of some of the best philosophers of the Twentieth Century could make a gallery of horrors.

Immanuel Kant is an exception. His life, seemingly uneventful, is inspired by the events of his day. It matches smoothly his thought, and the story of day-to-day happenings brings to life the intellectual and social world of his day. Kant's ideas develop linearly, and are best taught in a historical presentation such as Gulyga's.

Anyone concerned with problems of foundations, especially in the sciences, has to come terms with the philosophy of Immanuel Kant. Here as elsewhere, the price of ignorance is clumsy rediscovery and eventual ridicule. After a long night of the soul, scientists are again turning to philosophy. They will make a good start by reading Gulyga's biography, perhaps the

---

[1] Gulyga, *Immanuel Kant,* Birkhäuser, Boston, 1986

finest of Kant's biographies to-date. Gulyga's secret is the simple device of addressing a public of cultivated persons, rather than one of professional philosophers. This allows him to give details, philosophical and other, that a professional might find insultingly obvious. Gulyga's is an introduction to the philosophy of Immanuel Kant presented in the guise of biography, remarkably free of quirks and fads, even those of dialectical materialism.

Some day, after our civilization is gone, someone will draw a balance sheet of the philosophical heritage of the West. They will be looking for contributions of permanent value, for solid systems which in their time were able to provide, together with originality of philosophical insight, moral sustenance and a coherent view of Man and World. Very likely, only three movements will stand the test of time: the Academy of the Ancient World, the Scholastic of the Middle Ages, and the movement initiated by Kant, what came to be called German idealism, that is today the anchor of our thought and will remain so for the foreseeable future.

1985

CHAPTER TWENTY-FOUR

Heidegger

GENETIC LOGIC, the method of logical analysis inaugurated by Edmund Husserl, has open new and wide horizons in studies of the foundations of the sciences. Husserl's innovations, however, were rendered less effective and less acceptable by an injection of metaphysical views (as first observed by Nicolai Hartmann) resulting in obscurities and misunderstandings. Curiously, as Husserl's metaphysics stiffened with age, his logic was worked to a deep and pliable technique, and his later writings (*Formal and Transcendental Logic,* for example) are a blend of logical insights and questionable assertions on the transcendental ego and intersubjectivity. This blurring of the distinction between logic and metaphysics hardly makes for an attractive presentation, least of all to an outsider who has not been broken in to Husserl's manifold quirks. Genetic logic must be freed from all remnants of idealism if it is to become an autonomous discipline, as Husserl himself might have wished before 1929. The present note is meant as a step in this direction.

No object can be *given* without the intervention of a subject who exercises a selection of some features out of a potential infinite variety. It is impossible, for example, to recognize that three pennies placed side-by-side with three marbles are both "the same number," without focusing upon "the number of" and disregarding other similarities, for example, color, which may be more striking in other circumstances. This act of focusing, related as it may be to memories or triggered by associations, is logically *irreducible* to objectivity alone. It requires a contribution from the perceiving subject which, by its very selectivity, is ultimately arbitrary, or more precisely, *contingent.* We are thus forced to conclude that the constitution of every object in the world, as well as the structure of the world itself, depends upon human choice. In every such choice, no matter how well rooted historically or how far from haphazard, the possibility is ever lurking that the object may be or become wholly other than it is. The specter of random change hides in the shadow of man's every intervention. When this conclusion turns into a shocked realization, one obtains Heidegger's concept of *Angst.*

Yet, we know from experience that objects come furnished with the quality of already-there-ness. The world is constantly present in marmoreal facticity. The stars in the sky are not easily confused with chickens in a celestial coop, as Gracián's Iberian metaphor would have them.

We are confronted with a paradox. One the one hand, the necessity of an observer turns the object into a contingent event, whose existence hangs upon the thread of recognition. On the other, pre-givenness of the world confronts us as the most obvious of realities. Like all paradoxes, the paradox results from withholding part of the truth, and it melts away as soon as the motivation for each of the two alternatives is looked into.

A craftsman, a scientist, a man of the world regard the objects of daily commerce primarily as *tools.* It is a prerequisite of a tool that it offer its user a reliable guarantee of sameness. Only when the tool fails or becomes otherwise *problematic*, as when the need for a new tool is felt, does the contingency of all tool-ness reveal itself to the user, like a crack in a smoothly plastered wall. In such moments of *crisis*, the tool-user is forced to engage in a search for origins, for lost motivation, for forgotten mechanisms that had come to be taken for granted, much like a garage mechanic in his daily work. Thus, the two alternatives, necessity and contingency in the makeup of the object, far from being an irresoluble paradox, turn out to be complementary threads in the fabric of all organized endeavor. Genetic logic is the formal study of this dialectical structure.

Applied to the soft sciences, genetic logic is primarily the logic of concept-formation, designed to counter the trend towards uncritical adoption of notions of everyday life as structural concepts. In the life sciences, for example, concepts of current use such as "life," "evolution," or "organic," resemble, if only by their plebeian origins, the four Aristotelian elements. But chemistry was born only when four well-established notions out of everyday experience, such as water, fire, earth and air, were replaced by artificial concepts which were not at all to be found in experience, but were instead the result of an authentic genetic search, quite different from the naive recording of natural phenomena as they appear to the senses. Whether and when the life sciences will follow the example of chemistry and take a similar leap forward is a cliffhanger currently watched with anxiety in the scientific world.

In the exact sciences, the task of genetic logic is primarily the critique of foundations. Here, a proliferation of constructs, spurred on by unbelievable success — at least in the past — has led to a Babel of layered theories. Progress is made difficult by the weight of an awe-inspiring tradition. As Martin Heidegger admirably put it: *"Die hierbei zur Herrschaft kommende Tradition macht zunächst and zumeist das was sie "übergibt" so wenig zugänglich, dass sie es vielmehr verdeckt. Sie uberantwortet das überkommene der Selbstverständlichkeit und verlegt den Zugang zu den ursprünglichen "Quellen," daraus die überlieferten Kategorien und Begriffe z. T. in echter Weise geschöpft wurden. Die Tradition macht sogar eine solche Herkunft überhaupt vergessen. Sie bildet die Unbedurftigkeit aus, einen solchen Rückgang in seiner Notwendigkeit auch zu verstehen"* (Sein und Zeit, page 21).

The foundations of mathematics, as they appear today, faithfully mirror Heidegger's description. It is a law of nature that mathematical concepts shall be formed by a pattern of identification which must always erase its footsteps; as time and theory proceed, layers of usage, history, tradition and self-interest are heaped on, until the concept acquires a giddy air of always-having-been-around-ness. Uncovering the all-but-irretrievable origin is the purpose of genetic logic. Such uncovering requires a backward glance through an *ideal* time, not a historical glance through a *real* time. It calls for the reconstruction of an intentional, not a real history, a reconstruction which can only be carried out *in the light of* a problem at hand which acts as motivation. In mathematics, as elsewhere, all history is contemporary history, as Croce was wont to repeat.

At the turn of the century, the overhauling of the foundations of mathematics was entrusted to logicians, who soon found it eminently respectable to let foundational problems drift, and acquiesced as their discipline turned into another well-oiled cog in the mathematical machine, thereby losing all philosophical relevance. Genetic logic, which came along at approximately the same time, fits the problems of contemporary mathematics like a shoe, leading to the suspicion that all of Husserl's logic may have been secretly tested against a mathematical target. Mathematics played for Husserl an exemplary role in the hierarchy of learning; it was and still is the only successful instance of an eidetic science, and one that sets an example to be imitated, though not copied.

Consider for example the notion of *set*. It is intentionally related to the concept of *number* by what may be called the *problematization of identity*. Numbers had for centuries been taken as safe and sound working tools, but became problematic under the challenge of the notion of infinity ("Are infinite numbers possible?"). This forced the question "Where do numbers come from?" and called not for an historical, much less for a psychological or causal investigation, but for an intentional one. It is not easy to relive the feeling of triumph of the early logicians — near the turn of the century, that is — as they discovered that they could see through the opaque object "number" by introducing the function "the number of *x*." The new object *x* had to be named, since any function has to have a domain; hence the set. In this instance, genetic analysis revealed that every number results from an identification, namely, that it is an *equivalence class* (the name has stuck) of ontologically prior concepts, the newly uncovered sets.

The entire phenomenological theory of object-constitution

(lucidly developed by R. Sokolowski) is patterned after this typical — and admittedly oversimplified — scheme: from number to set via equivalence class. It is a central claim of genetic logic that *every* ideal object can be similarly analyzed — and must be, when it turns problematic — by displaying it as the result of a *recess* (*sit venia verbo*) onto a more fundamental object. To uncover the hidden origin of concept-object A is to display A as stemming from a prior concept-object B. Whether B historically pre-existed A, or whether it required an inventive act to come into being, is an idle question to the logician, though possibly not to the metaphysician.

The theory of sets set the pace and the standard for the technique of forging new mathematical concepts by shedding an irrelevant superstructure from old ones, widely used nowadays in mathematics. Take for instance the notion of topological space, at present taken for granted as if it always existed and as if it were to remain permanently with us, never again to change. Few students in our day would be ready to admit that it is an offspring of the far more sophisticated notion of algebraic surface. Yet, it all began with the question "Do we really need an algebraic formula to have a surface?" also asked around the turn of the century. It was a daring flight of fancy at the time to conceive the idea that perhaps surfaces could exist without the aid of a fixed set of coordinates, and without an ambient space to boot. It took yet another generation to realize that surfaces could be constructed even without the notion of distance, and somewhat later the requirement of a dimension was ditched. What was left became commonplace twenty years ago, and is now taught to undergraduates.

Had genetic logic been turned into a formal technique — and here it is a problem even to spell out what we ought to mean by *formal* — then it might now be endowed with the predictive ability that alone justifies a science. It could be used to spot the problematization of a scientific concept, and to prescribe the kind of intentional investigations that might lead to the discovery of an explanatory prior concept. For example, it is clear from several sources — physics, geometry, probability — that at the time of the present writing ominous clouds are gathering upon the concept of set, while strange and disparate new notions are being put forward as means to see through sets, backward to some fundamental pre-concept: Grothendieck topologies, topoi, modular spaces, etc. We may soon have to put up with pre-sets as our ancestors had to put up with pre-numbers, i.e. sets.

Whether or not genetic logic will develop into the first successful system of inductive logic depends on our ability to rework the programmatic writings of Heidegger, Hartmann, Husserl, Merleau-Ponty and Ortega into a discipline that can withstand the rigors of mathematical logic. The unparalleled scientific crisis of our time works in our favor. Seventy-two years after the publication of Husserl's *Logical Investigations*, the barbarians are besieging the citadel of phenomenology and crying in guttural accents: "Put up or shut up."

1977

CHAPTER TWENTY-FIVE

Doing Away with Science

THE VIEWS I AM ABOUT TO PRESENT are strictly my own. I rather fear that many of my colleagues and friends in the scientific community will strongly disagree both with my sentiments and with my conclusions.

I firmly believe that liberal arts colleges have an important and vital role to play because now more than in the past they are needed to counterbalance an ever-growing, and I may add potentially crippling, professionalism. Their failure to assume this responsibility will result not just in their own extinction but in the extinction of the whole concept of liberal education as well.

As I see it, the main problem which the colleges now face can be described as the "double squeeze." From below they are threatened by the high schools, which are reaching out for larger and larger chunks of the curriculum of the first two years, and from above by the graduate schools, which are encroaching to an ever-increasing degree upon the curriculum of the last two.

This is particularly true of mathematics, physics and chemistry because of unprecedented amounts of money and energy which have been poured out in recent years to strengthen education in these subjects both on the secondary level and, through the support of research, on the graduate school level.

A liberal arts college within a university with a strong graduate school can cope with this trend by succumbing to it. Hence the talk of cutting out the freshman year by admitting largely students with advanced standing, of the "six-year Ph.D. program" and other such schemes and proposals.

This trend is easily recognized as a trend toward the European system: secondary school followed by professional or graduate training — a system in which the liberal arts college as we know it is entirely eliminated. Let me now state the arguments of those who believe that this trend is both inevitable and good. The arguments are largely confined to mathematics and the sciences and they run roughly as follows.

There has been in recent years an explosion of knowledge. The frontiers of science have been pushed back (if you will pardon the cliché) with dazzling speed over fantastic distances. To join in this great adventure one must be in the thick of things, close to where new truths are found and close to those who are leading the search. To stand a chance at all, talented young men and women must jump in early, polarize strongly and run fast.

Smaller, independent liberal arts colleges, the argument continues, can neither attract gifted researchers to their faculties nor provide their students with adequate laboratory facilities and a sufficiently wide spectrum of specialized courses. A future mathematician or scientist, by choosing a small college rather than a large university, delays his entry into his chosen field and handicaps his future career.

The implications of this line of reasoning are clear; they are also extremely frightening. If we accept them we may well cause an irreversible change in the whole of our educational process on the basis of ill-digested and partly erroneous premises.

And in connection with the educational process it is well to recall a quotation from Falkland which was one of President Kennedy's favorites: "If it is not necessary to change then it is necessary not to change." How quickly one can reach absurdity can be illustrated as follows:

Assumption: We cannot attract first-rate people to the faculty unless we give them an opportunity to teach advanced, graduate courses.

Conclusion: We must create a graduate school of our own.

Result: A faculty of such excellence as to gain accreditation to engage in graduate education wasting its energy on third-rate graduate students.

For it is no secret that in mathematics, for example, an over-whelming majority of top-notch students (as judged by NSF or Woodrow Wilson awards) flock to only three or four schools. Leading schools like Columbia, Cornell, Stanford, Yale and many others, on whose faculties one finds some of the most gifted and creative mathematicians, complain — with good reason, I may add — that they do not get a fair share of top talent. What is ironic is that many of these top-notch students come from colleges which are now either contemplating or actually engaged in grafting on graduate schools.

Lest I may be misunderstood let me hasten to say now, as I shall repeatedly say in the sequel, that I am not against college faculties engaging in advanced teaching research and scholar-ship. On the contrary — they *must* engage in advanced teaching, research and scholarship. But I shudder at the thought of liberal arts colleges getting into the assembly line of what is called "Ph.D. production." They very term "Ph.D. production" is the saddest comment on what it supposedly represents.

Of course, the proponents of venturing into graduate educa-tion will be quick to point out that, unless opportunties for graduate work are made available, the colleges will soon be unable to attract undergraduates gifted in mathematics and the sciences. If this were true, then indeed independent liberal arts colleges would have to abandon science and mathematics, except perhaps as token lip service to the other of C.P. Snow's two cultures.

But the absurdity remains. For now the argument is that in order to attract gifted undergraduates one must also keep on hand a certain number of not-so-gifted graduate students so as to be able to attract gifted research-minded teachers who would then spend much of their energy teaching the less gifted ones. The resulting discrimination seems to be directed against stu-dents whose only fault is that they have not yet graduated from college.

I know that I am not being entirely fair to the other side but the picture I am painting though admittedly a caricature, is much closer to reality than one might be willing to admit.

So much for the trend. Let me now deal with the premises

on which the trend is based and try to demolish at least some of them.

The assumption easiest to deal with is that high schools can soon be entrusted with a significant share of the burden of providing a general education. Reference to the European experience, where essentially all general education is left to secondary schools, is irrelevant ad misleading.

European secondary schools are more selective than even our toughest colleges and only a very small proportion of young people achieve the baccalaureate. Teachers in secondary schools must all have a university degree with a training roughly equivalent to what we require of our Ph.D.'s except for a thesis. Before the second world war they were among the better paid civil servants and enjoyed far greater respect and status than American communities are willing or able to grant their high school teachers. Parental pressure was unthinkable, and the only way Johnny could escape six years of Latin and four years of Greek was to move to a different type of school (often in a different town) where, instead, he would have to do an equivalent amount of work in modern languages, history or biology.

Although enormous strides have been made in improvising high school education it will be at least 25 or, more likely, fifty years before one can begin to contemplate seriously, on the national scale, relegating a major portion of liberal education to our secondary schools.

More subtle and more pernicious is the implication that by not specializing at the earliest possible and allowable instant one is being sinfully wasteful. I particularly abhor this "efficiency expert" approach to education and I cannot resist a quotation even though it is taken from a review of a book on an advanced mathematical subject. "This book," wrote the reviewer, "is reminiscent of a bus tour through beautiful country where the purpose is not to admire the scenery but to keep on schedule."

The advocates of increased efficiency also point to the European system as a model of "wasteless" education. There, they say, once out of secondary school you are free to do only what you really want to do. If you are a mathematician, you can study only mathematics; a physicist, only physics, etc. No more of this nonsense of broadening one's outlook, of fulfilling irksome distribution requirements and coping with other "wasteful" impediments.

Well, I was educated in Europe. I entered the university (i.e. the equivalent of our graduate school) at the ripe age of seventeen, having been a few months earlier certified by the Polish Ministry of Education as being "mature to pursue higher studies."

To complete one's university studies and receive the lower degree (equivalent in our terms to completing the course work for a Ph.D.), one had, in addition to presenting a thesis (which need not contain original material) and passing an exam in a broad area related to the topic of the dissertation, to pass examinations in *eight subjects*. Of these only four were in mathematics proper. The remaining four were history of philosophy, experimental physics, rational mechanics and an elective which could be either logic, astronomy, crystallography or a branch of theoretical physics (exclusive of mechanics). Thus the efficient European system forced me and my contemporaries to "waste" roughly half of our time on "peripheral" subjects.

What about our American counterparts of today? More and more frequently we encounter holders of Ph.D. degrees in mathematics who have never heard of Newton's laws of motion and, to add insult to injury, a few who are even proud of it. At this time we are in danger of having so many "cultures" as to end up with no culture at all.

But the crux of the problem lies in the premise that the educational system is incapable of coping with the present explosion of knowledge and that therefore it has to be radically overhauled. This is a tough one and I shall have to deal with it in stages.

First, a glance at history. Between 1900 and 1910 the hitherto hypothetical atoms became a physical reality, Planck showed that energy can be emitted and absorbed in discrete quanta, and Einstein forever changed our outlook on space and time by his special theory of relativity.

Between 1910 and 1920 Niels Bohr discovered the first clue to atomic spectra and Einstein brought about a most profound merger of physics and geometry by showing the connection between the curvature of space and the force of gravity.

Between 1925 and 1940 the great edifice of quantum mechanics was erected and perfected, providing us with a deep and novel view of atomic and nuclear phenomena. Compared to these the present explosion of knowledge looks like a pretty tame affair.

True, in biological sciences there has been in recent years an explosion of monumental proportions. But, as far as I can tell, my colleagues at the Rockefeller University manage to stay remarkably calm. I have not yet heard anyone seriously advocate that biochemistry be taught in high school and, while many of our students in biological sciences are, upon entering, deficient in biochemistry, the problem is dealt with by an intensive and stiff summer course.

In the frantic years after the second world war we have somehow lost sight of the fact that, to quote James Stephens, one can know less but understand more. Education cannot be based on being frightened by the immensity of knowledge but on the promise of insight which comes from understanding and mastery of basic principles. To deny this is to deny every lesson of history and to reject much of what shaped our civilization.

Professional education concentrates on *skills*. Liberal education must concentrate on *attitudes*. I am not trying to minimize the importance of skills. That they are necessary goes without saying. But it is not necessary, and in fact it is foolish, to attempt to teach all the skills that are currently in use in this or that branch of science. It is infinitely preferable to concentrate on developing in students courage to tackle a problem and willingness to face anything they may encounter. But to instill this kind of courage the teachers themselves must have it.

And here we come to the toughest problem of all: Who? Who will and who can do it? The growing professionalism and specialization in graduate schools have contributed to an atmosphere in which teaching and research are somehow pitted against each other. (The status symbol now, for example, is not measured so much in dollars per year as in hours per week *not* spent in the classroom.) Coupled with this there is a genuine shortage of adequately trained mathematicians and scientists to fill the manifold needs of our society and, in particular, a great shortage of those who can be attracted into college or university teaching.

We have all heard reports of neglect of undergraduate teaching in large universities. Some of these reports are particularly frightening because they seem to suggest a lack of concern bordering on irresponsibility. It would, however, be a serious mistake to draw the conclusion that those interested in research are notoriously bad and indifferent teachers, and consequently that to be a good teacher one must abandon whatever commitment to research one may have.

Let me be frank. A benevolent Mr. Chips who does nothing but teach freshmen calculus and hold the hand of every homesick student is as inadequate a teacher as the impatient, brilliant young expert in algebraic functions or functional analysis who looks upon his teaching of freshman calculus as an indignity and a bore. Fortunately, both Mr. Chips and B. Y. E. (brilliant young expert) are largely non-existent in the flesh — stereotypes which each side uses to frighten the other.

The two sides I am referring to are, on the one hand, those who fear that emphasis on teaching may be used as a façade to hide mediocrity and ignorance, and on the other, those who fear that capricious and socially irresponsible people are hiding behind it ever-popular banner of research.

At the bottom of it all lies an unfortunate truth. It is that a not insignificant number of those who go into teaching in small liberal arts colleges slip into devoting more and more of their time to routine courses and making themselves generally useful through advising, committee work and other chores. They do not keep up with their subject and they soon are neither willing nor, I am sorry to say, able to teach anything beyond the most elementary material. A college which has allowed its science faculty to reach this level has in effect abandoned science.

Let me then state it as an axiom that in order to be a good teacher, not just a popular one, one must have such an unwavering dedication and such an unbreakable commitment to one's subject that intellectual somnambolism becomes unthinkable. But it is equally axiomatic that to be a good teacher one must be capable of joy and satisfaction in passing on the torch even if other hands made the fire that lit it.

Render unto Caesar the things which are Caesar's.

Liberal arts colleges will not attract to their faculties, except in rare instances, the most brilliant, the most original or the most creative. And perhaps that is just as well. But there are other kinds of creativity which are not measured in terms of great new discoveries or deep penetrations into the unknown.

I once had the pleasure of meeting the late Francis Friedman of M.I.T. whose premature death a few years ago is still mourned by all who knew him. He was one of the moving spirits behind the new, highly imaginative high school physics course (P.S.S.C.), and I heard him discuss the course and show some filmed experiments which he and others had designed. It was a remarkable performance, and for a moment I felt the kind of elation I had not felt since I was a very young man and was first introduced to the wonders of science.

My friend Martin Klein, professor of physics at Case Institute of Technology, who made many signficant original contributions to physics, became interested in the history of the period when the idea of quanta first made its appearance. His brilliant and fascinating analysis of Planck's discovery is another example of the kind of creativity that may easily go unrecognized if one applies the "professional" standards so prevalent today.

There are many more potential scholars-teachers of the kind I have tried to describe than most people think. In my own limited experience I have come across quite a few. The trouble is that many of these gifted people consider it a sign of some kind of failure to devote their lives to teaching in liberal arts colleges unconnected with a graduate school or otherwise removed from centers of research. Here we are dealing with a perversion of standards which it may take a generation or longer to cure.

In the meantime, to attract at least some of the available talent one can begin by taking a few steps which, though obvious, may nevertheless bear stating.

1. An admission policy aimed at selecting those with *genuinely* strong and wide cultural interests and with a commitment to an *area* of knowledge rather than to a specific subject. A seventeen-year-old who is absolutely bent on becoming an organic chemist should probably seek his fame and fortune elsewhere. A youngster whose interests are in science but who is not in a hurry to specialize is a better candidate, even if his College Board scores are lower and his all-round record spottier.

2. More imaginative ways and means of teaching. Examples of this are the Honors Program at Swarthmore and the College Plan at Wesleyan.

3. Encouragement of seminars and courses taught jointly by members of different departments: logic by a mathematician and a philosopher, differential equations by a mathematician and a physicist, organic chemistry by a chemist and a biologist, etc.

4. Opportunities for teaching advanced courses and encouragement of seminars in which the whole staff will participate.

5. Inviting visitors from great centers of research and letting them participate with your own staffs in teaching for a signficant period of time (at least two weeks).

6. Cooperating closely with professional societies in their efforts to organize summer seminars designed to bring members of college faculties up to date in at least some areas of mathematics and science in which lively research is currently being done.

7. Be on the lookout for scientists whose interests turn to the history of science (there are more and more of them). Grab them and let them teach both science and history.

8. Finally, apply constant pressure on graduate schools by reminding them that it is their responsibility to train teachers and not just experts in this or that. Students, who go on to graduate school could form the nucleus of a lobby for a more balanced, more reflective and less feverishly professional education.

While the explosion of knowledge has, in my opinion, been greatly exaggerated, the explosion in the number of practitioners of science has indeed been staggering. It is the latter explosion that has radically changed the appearance of science and set it apart from its traditional setting.

Some of the results have been unquestionably beneficial. But the price has been high and may go even higher. I am not just thinking of the invention of nuclear weapons, though this alone is quite a price to pay for progress. Since I must assume that we will somehow escape nuclear destruction, I am thinking more of the price we are paying in deterioration of critical judgment, in severance of lines of communication between closely related disciplines and, last but not least, in increasing cultural isolation.

In a way, scientists themselves have abandoned science by allowing it to become a set of professions rather than one grand rational scheme of perceiving the universe. That is why I plead with liberal arts colleges not to abandon science, and I can only hope that my plea does not come too late.

1981

---

[1] I may add that there is some inexplicable tendency to equate *advanced* with *graduate*. I should like therefore to mention that probably the most advanced, certainly the second most advanced, course I taught during my 22 years at Cornell was an honors seminar for a small group of upperclassmen.

CHAPTER TWENTY-SIX

More Discrete Thoughts

THE COMPUTER IS JUST AN INSTRUMENT for doing faster what we already know how to do slower. All pretension to computer intelligence and paradise-tomorrow promises should be toned down before the public turns away in disgust. And if that should happen, our civilization might not survive.

Unfortunately, the reigning mediocrity of our day demands a heavy toll of any scientist who dares stray from routine publication of technicalities, over into the blue yonder of speculation. Even one of the great physicists of our time used to pan younger upstarts with the putdown: "X is prematurely wise."

It was a goof of the late John Stuart Mill, the incorrigible child of philosophy, to place the term "inductive logic" on a par with the soundly established "deductive logic," and to think of it as a sort of deductive logic with a negative sign. Since then, the idea has fallen into the hands of professional oversimplifiers, bent on finding the laws of induction, cost what may, in parallel with the laws governing the syllogism and the existential quantifier. The inductive logicians are assuming the role of Einsteins in a subject that never had its Galileo. Why don't we tell the truth? No one has the faintest idea of how process of scientific induction works, and in calling it a "process" we may be already making a dangerous assumption.

The fumbling first attempts at an effective theory of perception should have a salutary effect in the dusty halls where philosophers are wont to camp. Questions that are used to be purely academic, and thus unverifiable, are now suddenly in the forefront of science, thanks to the computer, and philosophers who have never thought they would be called to acccount for their speculations — at least in this life — find themselves in an uneasy spotlight. At the other end, engineers that never thought they would have to read another humanistic book now look for the help and solace of speculative Kultur.

Unlike most historians, J.F. Hofmann writes engagingly and accessibly. Leibniz in Paris should go a long way to do away with the perniciously inaccurate Romantic image of the superior-to-all, universal, saintly "genius," an image which is still inculcated, with criminal disregard for the truth and catastrophic results, to schoolchildren all over the world.

How much longer will the present *folie* for precision in philosophy last? Need a concept be precise in order to be meaningful and effective? Or do philosophers wish to commit hara-kari on the altar of mathematics?

When a philosopher writes well one can forgive him anything, even being an analytic philosopher.